POSTHUMAN BODIES
KA 0160705 7

Unnatural Acts:
Theorizing the Performative

Sue-Ellen Case
Philip Brett
Susan Leigh Foster

The partitioning of performance into obligatory appearances and strict disallowances is a complex social code assumed to be "natural" until recent notions of performativity unmasked its operations. Performance partitions, strictly enforced within traditional conceptions of the arts, foreground the gestures of the dancer, but ignore those of the orchestra player, assign significance to the elocution of the actor, but not to the utterances of the audience. The critical notion of performativity both reveals these partitions as unnatural and opens the way for the consideration of all cultural intercourse as performance. It also exposes the compulsory nature of some orders of performance. The oppressive requirements of systems that organize gender and sexual practices mark who may wear the dress and who may perform the kiss. Further, the fashion of the dress and the colorizing of the skin that dons it are disciplined by systems of class and "race." These cultural performances are critical sites for study.

The series Unnatural Acts encourages further interrogations of all varieties of performance both in the traditional sense of the term and from the broader perspective provided by performativity.

Posthuman Bodies

edited by Judith Halberstam
and Ira Livingston

INDIANA UNIVERSITY PRESS

Bloomington and Indianapolis

The paper used in this publication
meets the minimum requirements of American National Standard
for Information Sciences—Permanence of Paper
for Printed Library Materials,
ANSI Z39.48-1984.
∞™

Manufactured in the United States of America

Library of Congress Cataloging-in-Publication Data

Posthuman bodies / edited by Judith Halberstam and Ira Livingston.
p. cm.—(Unnatural acts)
Includes index.
Romanized record.
ISBN 0-253-32894-2 (alk. paper).—ISBN 0-253-20970-6 (pbk. : alk. paper)
1. Body, Human—Social aspects. 2. Body, Human—Symbolic aspects.
3. Body, Human (Philosophy) 4. Body image. 5. Sex symbolism.
6. Sexuality in popular culture. 7. Humanism—20th century.
8. Postmodernism. I. Halberstam, Judith, date.
II. Livingston, Ira, date. III. Series.
GT495.P67 1995
391'.6—dc20 94-45934

1 2 3 4 5 00 99 98 97 96 95

Contents

PART III

Queering

PART IV

Terminal Bodies

Preface

Posthuman Bodies is a collection of essays that takes up, in a mostly affirmative way, various challenges to the coherence of the "human body" as a figure through which culture is processed and oriented. In the essay that opens this volume, we argue that a posthuman condition is upon us, and that nostalgia for a humanist philosophy of self and other, human and alien, normal and queer is merely the echo of a battle that has already taken place. This argument is not a truth claim but, like the title of the volume itself, an open invitation to engage discursive and bodily configurations that displace the human, humanism, and the humanities. As we will assert, such engagements come from the experience that the authorization that these identities seem to offer comes at too high a price; at the price of rendering unintelligible much of what matters to us. The "us" of this pronouncement and what is at stake in it will be constantly under revision in the essays that follow.

Like "us," most of the contributors to this volume come *through* the humanities—diagonally, as it were—neither quite beginning there, nor quite leaving them behind. The essays cluster around film and literary studies, cultural studies of science and science fiction, feminist and queer studies, but multiple other resonances and disjunctions characterize relations within and among volume, sections, and essays. Typically, these do not yield the glimpse of some utopian interdisciplinary space or Program in the Posthumanities; instead, they share a commitment in practice to hybridities that resist reduction to single principles; a perversity that is often enacted through diagonal resistances to standard academic discourse. While this may seem an annoyance or even a failure of organizational rigor to some readers, it is for better and worse a primary animator of the assemblage of *Posthuman Bodies*. To put it another way, the "post" of "posthuman" interests us not really insofar as it posits some subsequent developmental state, but as it collapses into *sub-*, *inter-*, *infra-*, *trans-*, *pre-*, *anti-*.

Following our manifesto-manqué, Allucquere Rosanne Stone and Steven Shaviro elaborate some of the multiplications that characterize the posthuman condition. In "Identity in Oshkosh," Stone explores how Multiple Personality Disorder, which she follows through a Wisconsin rape trial, represents a crisis in accountability and agency, raising potent questions about "how cultural meaning is constructed in relation to bodies and selves." Shaviro suggests in the following essay that we can learn a couple of things from William Burroughs about the multiple and parasitical relations between bodies and personalities, words and money, insects and sexual torture. For Shaviro as for Burroughs, "Self-identity is ultimately a symptom of parasitic invasion, the expression within me of forces originating from the outside."

If the lessons from Burroughs and the enterprise of multiplicity still seem unclear, Kathy Acker, in her inimitable bad-girl vogue, clears up the matter: "It was the days when men were cutting off their cocks and women were putting on strap-ons." Acker's apocalyptic soap opera continues to take us through masochism, paranoia and plagiarized sex tips for girls, assembling along the way an anti-Oedipal sex/gender system whose "someness" always makes it the site of conflicting stories-in-process. The three following essays explore the someness of gender for the opportunities it offers for explicitly feminist engagements with posthumanity. Alexandra Chasin considers how "Identities among Women, Servants, and Machines" (and subject/object distinctions historically bound up with them) are renegotiated in various human-machine interactions, suggesting that human and machine "working beings" share a common project in undermining claims of ontological difference between us. Paula Rabinowitz exploits the equally shifty boundaries between documentary truth, cinematic fantasy, and spectatorial investment in Chick Strand's film, *Soft Fiction*. Susan M. Squier moves the performance space from the dark comforts of the cinema to the intrauterine space of reproductive technologies, tracing the emergence of a new constellation of reproductive images—"Ectogenetic Fetus, Surrogate Mother, Pregnant Man"—to assess their disciplinary functions in posthuman reproduction.

Essays by Jennifer Terry, Camilla Griggers and Roddey Reid are clustered around discursive practices of "Queering." Terry investigates how scientific discourses of sexual normalcy and deviance can be implicated in the construction of homophobia while retaining a "seductive power" for some gay men and lesbians. Griggers, on the other hand, is interested in how the mass media's image of the deviant lesbian body functions as a definitive limit case for postmodern and posthuman technologies of violent subjectivity. From the death of a pacifist femininity, Reid shifts the scene to the "Death of the Family" as a narrative that has both succeeded and failed in "Keeping Human Beings Human." Reid inverts the scare tactics of conservative discourse on the family, parodying the gothic imagery of decline and apocalypse to situate death-of-the-family narratives into a posthuman context.

The essays that comprise the final section, "Terminal Bodies," emphasize cinematic posthuman becomings in which "Terminal" signals not a simple end but termination as an ongoing condition, as well as indicating the place where human and machine interface, with unpredictable results. Kelly Hurley's "Reading Like an Alien" posits alternatives to psychoideological theories of the horror film that understand its function as the negotiation of cultural repressions. Hurley argues that, while the horror film works within traditional cultural narratives of "the human," including those of psychoanalysis, it does so in order to rupture and exceed them, generating new images and narratives of "the human" as posthuman in the process. In "Terminating Bodies: Toward a Cyborg History of Abortion," Carol Mason challenges the messianic strain in recent theoretical constructions of the cyborg as a symbol for political embodiment, arguing that historic divisions between class, race, and gender enacted in *Terminator II*—especially the film's apparent celebration of white femininity at the expense of black masculinity—compromise the liberatory potential of the cyborg. Eric White's "Evolutionist Cinema" tends to posit bodily futures as animations of "hitherto latent aspects of human nature." White traces several film narratives in which the human body becomes monstrously other by emancipating "the menagerie within."

We hope this volume will function as a kind of upgrade; a piece of very soft-ware that enacts and enables various interactivities, code-switchings, and other potentially viral discursive involvements. While we wait for a self-help book to tell us how to get in touch with "the menagerie within," this ambiguously pregnant set of cautions and exhortations will have to stand in for a user's manual. Go forth and multiply. Your actual mileage may vary. Objects in mirror are closer than they appear.

Judith Halberstam
and Ira Livingston

Acknowledgments

Our thanks go to Kelly Hurley and Eric White for their part in conceptualizing this volume, and to Sue-Ellen Case, Susan Foster, Philip Brett, and others at Indiana University Press for helping to materialize it. For granting some of the time and space to work on this project, Ira Livingston thanks the director, Peter Copek, and staff of the Oregon State University Humanities Center. We thank Verso Press for permission to publish Paula Rabinowitz's essay, a version of which appeared in her book on documentary film. We thank Grove Press for permission to publish Kathy Acker's essay, "The End of the World of White Men," a subsequent version of which appears as the preface to *Pussy, King of the Pirates*, under the title, "Once Upon a Time, Not Long Ago, O."

POSTHUMAN BODIES

Introduction: Posthuman Bodies

Judith Halberstam and Ira Livingston

> If the time should ever come when what is now called science, thus familiarised to men, shall be ready to put on, as it were, a form of flesh and blood, the Poet will lend his divine spirit to aid the transfiguration, and will welcome the Being thus produced, as a dear and genuine inmate of the household of man. (Wordsworth 738)

Now that Wordsworth's entrepreneurial speculation of future collusion between scientific and cultural production has paid off repeatedly, the bond matured and the stock split and reinvested again and again to the profit of its stockholders, the loyalty of employees and customers of the human monopoly (Nature/Culture Systems, Incorporated) can no longer be assured. Science and its poetic sidekick have maintained the "household of man" through exclusions, subordinations, exoticizations, pathologizations, criminalizations—thus guaranteeing that the "transfiguration" that is upon us cannot leave intact any of Wordsworth's interdependent terms: neither "what is now called science," nor the "form of flesh and blood," nor the "household of man."

Posthumanities emerge not in the happy interdisciplinary family business imagined by Wordsworth, but (equipped with leaked secrets and embezzled powers) out of a disenchantment that is both anti-aesthetic and anti-scientific. It is in this volatile market that the medical/aesthetic disciplinary monopoly on "the body" is being challenged. If the announcement of the discovery that "the body" has a *history* has become conventional, the field that it inaugurates has only begun to be established. Even so, the emergence of "the body" in history, and thereby its partial reification and relativization, also opens a space for posthistorical bodies to establish themselves.

"We're all connected," crooned a recent ad-campaign for New York Telephone: that was the kind of thing Wordsworth had in mind. The slogan performs an exemplary ideo/topo-logical maneuver. The organicist notion of connectedness—and its most extreme mystification, the Romantic imagination—had been invented as internalizations and de-politicizations of dominant material interests and their power/knowledge grid. The ad turns the heavily laundered Romantic imagination inside out to organicize the corporate body. The old humanist party line is sublated in the postmodern partyline, dogma mutated into a floating multiple conversation, couplings into switchboards—looking forward to an operatorless networking that is both and neither perfect freedom and the perfect police state (which, as William Burroughs reminds us, has no need for police). But if the extension, attenuation, miniaturization, and cross-wired interdependence of the networks that implicate the body are Control strategies (and they are), the time has passed when resistance could effectively be imagined in terms of a sovereign, local, man's-home-is-his-castle body. The price of indulging nostalgia for the immediacy of edenic nakedness, or for the spontaneous and bodily unity of the revolutionary crowd, is too high. The urgency for new kinds of coitions and coalitions is too compelling in an age of continuous and obligatory diasporas.

The constructionist body is not equal to the task if it is merely a compensatory or reactionary opponent to the humanist body. The proletarianization or automatization of the body with respect to "discursivity" is an anxious reaction-formation to the "loss" of an autonomy that was itself an exclusive fiction. Posthuman bodies are not slaves to masterdiscourses but emerge at nodes where bodies, bodies of discourse, and discourses of bodies intersect to foreclose any easy distinction between actor and stage, between sender/receiver, channel, code, message, context. Posthuman embodiment, like Haraway's "feminist embodiment, then, is not about fixed location in a reified body, female or otherwise, but about nodes in fields, inflections in orientations. . . . Embodiment is significant prosthesis" (195).

Sign Posts: Some Posthuman Narratives

Postmodernism, poststructuralism, postcolonialism, postindustrial capitalism: the proliferation of academic "post-isms" marks simultaneously the necessary or regrettable failure to imagine what's next and the recognition that it must always appear as "the as yet unnamable which is proclaiming itself and which can do so, as is necessary whenever a birth is in the offing, only under the species of the non-species, in the formless, mute, infant and terrifying form of monstrosity" (Derrida 293). But the rough beast that now slouches towards the next century is not monstrous simply by virtue of its status as a non-species: posthuman monstrosity and its bodily forms are recognizable because they occupy the overlap between the now and the then, the here and the always: the annunciation of posthumanity is always both premature and old news.

Posthuman bodies are the causes and effects of postmodern relations of power and pleasure, virtuality and reality, sex and its consequences. The posthuman body is a technology, a screen, a projected image; it is a body under the sign of AIDS, a contaminated body, a deadly body, a techno-body; it is, as we shall see, a queer body. The human body itself is no longer part of "the family of man" but of a zoo of posthumanities. In their recent world tour, the rock group U2 coined the concept "Zoo TV" and performed the becoming-posthuman of the body on stage and on camera, somewhere between desire and captivity. Zoo TV was a remarkable performance of identity in mass media culture for several reasons. Bono's various couplings on stage with mirrors, cameras and video equipment fundamentally undermined otherwise stable relationships between fan and star, disconcerting the technology of rock stardom by insisting that the star is a trick of the dazzling lights, a feedback effect rather than an emotional center that anchors the rock performance in time and space for each individual fan.

Is the performer screen or image, reflection or production? By calling the rock extravaganza "Zoo TV," U2 confuses the distinction between who is looking out or in, who is in the cage, who looks on, who is exoticized, what is rare, who is catalogued and how. We might ask how Zoo TV collapses nature and culture into each other, into a place where captivity refers to a state of desire (fan captivation) rather than a state of siege. But is captivity on screen or off?

The relation between the posthuman and the postmodern in a Zoo TV society relies on a new technological order with the body at its helm and a troubling relationship to history. Speed and its possibilities—the speed of the new, the speeds of potential futures colliding with the fast approaching past—create a crisis in the category of "history" and the narratives it inspires. History is inefficient as a method of processing meaning; it cannot keep up. As history slows down relative to events in the realm of information and meaning, the future remains on hold. History as social or chronological history is dying with the white male of western metaphysics and consequently it is no longer enough to say where we have been. We struggle instead to articulate a present laden with the debris of inert pasts. Posthuman bodies do not belong to linear history. They are of the past and future lived as present crisis. This present, this crisis does not glide smoothly along a one-dimensional timeline but erupts or coalesces non-locally across an only partially temporizable realm of meaning.

Posthuman Bodies represents attempts to keep up with the present and to process the identities that rub up against the body and then dissolve in the maelstrom we call postmodernism, posthumanism, poststructuralism, postcolonialism, postindustrial capitalism. The essays in this volume work to engage posthuman narratives that have all but replaced previous masternarratives about humanity, its bodies, its subjects, its pains, and its pleasures. These narratives show how the body and its effects have been thoroughly re-imagined through an infra-disciplinary interrogation of human identity and its attendant ideologies.

Out Posts: Some Subcultures without Culture: *Paris Is Burning*

Posthuman bodies thrive in subcultures without culture: there are only subcultures. Culture processes and appropriates a subculture only as quickly as the subculture becomes visible as culture: the Imaginary of dominant culture is always only a culmination of appropriated forms and plagiarized lyrics (if a mirror can be said to appropriate anything).

Voguing, now a famous instance of the signifying dance of the hyperstylized body, began as a predominantly black and latino transvestite subcultural denaturalization of haute-culture gender performance (before being mainstreamed by a very white Madonna). But to identify voguing as parasitical on Big Culture (e.g., under the heading of "parody") would be as reductive as to try to understand voguing as Romantic Creativity. Instead, voguing and other subcultural practices work to undermine the one-eyed pyramid of generic hierarchy, to trouble the smooth flowchart of cultural circulation, somewhat like films that precede novelizations, sequels that precede prequels, mafia bosses that model themselves on movie mafia bosses, actor-presidents, TV-doctors who endorse pills, polls that pit sitting vice presidents against the TV characters they denounce, infomercials, docudramas, and so on and on.

Madonna mimics black and latino gay prostitute culture and translates it into a million-dollar stage act; her performances are attempts to originate the forms she has appropriated. This is exactly the process by which some performances are given the weight and authority of "reality" while others are relegated to shadows and imitations. But if authority and origination are conferred by the circulation of capital, this circulation can never quite establish its priority over the counter-fluencies of subculture.

Judith Butler exposes the relationship—between simulacrum and "original"—that troubles gender and compulsory heterosexuality. By inverting the dominant narrative of the relation between heterosexual gender performance and butch-femme lesbian gender performances, Butler is able to claim that "the parodic or imitative effect of gay identities works neither to copy nor to emulate heterosexuality, but rather, to expose heterosexuality as an incessant and *panicked* imitation of its own naturalized idealization" (22–23). This inversion is powerful because of the way it intervenes in the construction of gendered subjectivity at the point where it becomes a model of humanness. It interrupts a linear continuity among gender, heterosexual norms, and human sexuality by showing how heavily heterosexuality and gender depend on gay identities to idealize, humanize and naturalize their own definitions. This dependence is too often left out of accounts of the "Other" that stress marginalization. While clear and present oppression of "Others" is by no means to be understated, the Other is also the matrix against which the self is made to appear and from which it can never be extricated; the "conservation of Otherness" dictates that any "assimilation" or "incorporation" will also be a transfiguration.

Madonna inverts the relation between subculture and culture in a rather similar way to give the illusion of a monolithic culture of white monied heterosexuality, under whose camera eye she squirms. The release of Jennie Livingston's film, *Paris Is Burning,* drained Madonna's voguing extravaganza of its reality effect even while being pulled part way up by her bootstraps. Not only do New York City's drag queens give an alternative history to the origin of voguing, they also give an alternative history of gender and its performances. It is worth looking closely at this film in order to engage the posthuman narratives that saturate transitions between cultures and subcultures.

Balls, houses, legends; reading, throwing shade, walking; realness, categories, vogue: the subcultural "dictionary" that organizes *Paris Is Burning* insists on thoroughgoing rearticulations. At a ball, reality itself is up for grabs—and may the best queen win. Voguing, one drag queen explains, is a form of street-fighting; a competition waged between two houses or gay gangs. Houses are like families and they take their names from designers (House of Chanel, Saint-Laurent, etc.) or from their founders or Mothers (House of Labeija, Ninja, etc.). Between "tribe" and "family" and "profession" and "commune" and "corporation," the House is an unromanticizably opportunistic posthuman assemblage that could never be mistaken for the cozy privacy of Wordsworth's "household of man."

To "walk the ball" is to compete, in one of a huge range of categories, against "Children" of other Houses for trophies. Categories include Butch and Femme Queens, Realness, Bangee Girl and Boy, and so on. In Realness, Children simulate a social role to the point where they could pass for real. For example, Executive Realness involves dressing as a businessman with suit, tie and attaché case. The Realness category allows poor, gay, often black or latino men to untangle for a moment the economic and social forms of oppression that stand between them and the so-called "real world." It also allows them, however, to recreate that real world in their own image, to repeople it and to challenge in an intensely artistic way the conventions of domination.

While many of the Femme Queens are satisfied to strike poses of femininity, others in the ball scene have had actual transsexual operations. Bodily operations suggest that "Realness" may in fact have something to do with physical organs, while the drag shows suggest that, on the contrary, the most Real woman is one who passes on the streets rather than between the sheets. This tension between "real" anatomy and real gender is articulated by several Femme Queens in the documentary. Pepper Labeija and Dorian Corey offer accounts of what they perceive to be the nuanced distance between performing realness and wanting to be real. Corey says that the Children hunger too much for something beyond the "small fame" of walking the ball. Labeija cautions against taking realness for real; he never wanted the operation because he knows that simply "having a pussy does not mean you will have a fabulous life." Labeija wryly implies that becoming a woman means facing a new oppression: to be a "real" woman is simply to face "real" sexism. On the other hand, Venus Xtravaganza wants the operation and longs to be "a spoiled, rich, white girl living in the suburbs." While this kind of sentiment drew horrified responses from some liberal critics who marveled at the willingness of people to embrace their oppressions, it is a fantasy that actually begs to be read within the context of the balls and their codes of signification. Venus's fantasy functions as fantasy precisely because its realization will always be frustrated. The "real" of her fantasy, of course, has little if anything to do with spoiled, white girls in suburbs. The posthuman element of this fantasy lies in its non-relation to real whiteness and its expression of the fantasy function of white realness. Whiteness, in other words, functions in this fantasy as a *limit* of the real and as a desired category only because it is unattainable or impossible. Not because whiteness cannot be simulated but because Venus Xtravaganza for one will never reap the rewards of even a successful simulation of whiteness. Real whiteness, however, the other end of this equation, becomes equally *vulnerable* insofar as

Venus's fantasy makes visible the lines of power that collide in the category "white" and which allow it to slide into the category "human."

Madonna performs the real "whiteness" that voguing exposes as drag in order to stabilize the categories and make her whiteness and realness work for her in a way that Venus never can. While Venus and the other queens imitate a whiteness they find in fashion magazines, Madonna imitates the imitation in order to reclaim and re-secure voguing for superstars. Madonna's performance and her blond translation of voguing make her a real millionaire; Venus dies before the film project is completed, a murder victim. This is not, therefore, merely a moral lesson about the dangers of thinking realness is mutable. Instead, Madonna and Venus are examples of the power of stable real whiteness versus the risk and insecurities of trying to perform white realness. These are not aberrations of the flow-vectors that define the structure of cultural space-time but indicators of the poverty of teleological and hierarchical narratives to account for cultural traffic.

The gridlock of signifiers and signifieds at the juncture of gender, class, ethnicity, and sexuality in the night world of voguing is a traffic jam of posthuman proportions, where the drivers may as well abandon their vehicles. The Human wanders, lost, into a maze of sex changes, wardrobe changes, make-overs, and cover versions that imbricate human reality into posthuman realness.

As definitions of bodies and their acts proliferate within subcultures, they shrink proportionately in what we call culture. One example, *Husbands and Wives,* one of Woody Allen's melancholic autobiographical confessions, registers the loss of a sexual vocabulary within normative heterosexuality. Judy Davis plays a frustrated and frigid divorcée who struggles to find the right sexual combination, the formula she hopes will unlock her desire. After a date with a caring, handsome man she seems to like but not desire, she is reduced to thinking of coupling as the union of "hedgehogs and foxes," a union that signifies the impossibility of complementarity. But the model for binary complementarity in Allen's film is a heterosexuality that here seems stuck always in a mode of either/or, with no alternatives in sight. Davis's character lacks a way of understanding the desire for difference and the desire for sameness; where they overlap, where they collide, where they come to blows. Hedgehogs and foxes? Meanwhile, minority sexual cultures generate elaborate and proliferating sexual vocabularies: so many words, so many acts, so few discrete identities—or only as many identities as there are bodies and then some. Hedgehogs and foxes? This definition registers the pathos of normative heterosexuality locked into a sad groove, constantly generating narratives of sentiment and romance to cover over the obvious confusion and lack of faith that plagues all attempts to mate for life.

Someness

Sex only has currency when it becomes a channel for something besides its own drive for pleasure. Turn-ons are not sexual; sexuality is a dispersed relation between bodies and things: some bodies (such as male lesbians, female cockwearers, baby butches, generationalists, sadofetishists, women with guns) and some things (dildoes, pistols, vegetables, ATM cards, computers, phones, books, phone books). Some turn-ons: women in suits looking like boys, women in suits wearing dildoes looking like and being men, men without dicks, dicks without men, virtual body parts, interactive fantasy. What is bodily about sex? What is sexual about sex? What is gendered? Are posthuman bodies postgender? Is anything post anymore, or is this the beginning? The search for origins stops here because we are the origins at which imagined reality, virtual reality, gothic reality are all up for grabs. You're not human until you're posthuman. You were never human.

What would happen if singularities ceased to anchor the ways in which we think? Not The Posthuman Body, but bodies. "The sex which is not one" is the plural paradigm for the species which are never one. Deleuze and Guattari revise the paradigm of the subject strung like a marionette to reduce the marionette body and the puppeteer mind to more cat's cradles of nervous fibers, sets of intersecting bio-psycho-social constraints that make the nodal body (8). This is not to replace a stuck mind-body dualism with a heterogeneous monism, but to insist on the "someness" of every assemblage. Posthumanity cannot be asserted by a kind of gender suffrage (each person their own gender) because the discourse of "infinite diversity" just plays the "good cop" to the "bad cop" of singularity and duality, to the tendency to set up one (system of gender) and two (m/f, gay/straight, gay/lesbian). For Haraway's "cyborg," "one is too few, but two are too many" (177); Homi Bhabha's postcolonial "hybridity" is "less than one and double" (179); Deleuze and Guattari's "assemblage" is enumerated as "n minus one":

> In truth, it is not enough to say, 'Long live the multiple,' difficult as it is to raise that cry. No typographical, lexical, or even syntactical cleverness is enough to make it heard. The multiple *must be made,* not by always adding a higher dimension, but rather in the simplest of ways, by dint of sobriety, with the number of dimensions one already has available—always n minus one. (6)

How many races, genders, sexualities are there? Some. How many are you? Some. "Some" is not an indefinite number awaiting a more accurate measurement, but a rigorous theoretical mandate whose specification, necessary as it is (since "the multiple *must be made*"), is neither numerable nor, in the common sense, innumerable.

Some Humans

The rhetorical crisis for the humanist is such that one minute he'll lay down the law of the jungle to you and the next minute he'll be aghast when everything isn't tastefulness, gentility, and rationality. The privilege of blindness to these contradictions is part of the arrogance of entrenched power; no doubt it will always be ready to sacrifice everything, beginning as usual with its subalterns, in order to go to the grave with the privilege of this blindness, with the delusion of its own disinterestedness or internal consistency, the proud fiction of its self-sacrificing fatherliness or motherliness.

The posthuman marks a solidarity between disenchanted liberal subjects and those who were always-already disenchanted, those who seek to betray identities that legitimize or de-legitimize them at too high a cost. No one comes naturally to this conjuncture; rather it must be continually forged within and among people and discourses.

When Air Force pilot George Bush dropped his bombload on his target and bailed out, regrettably but unavoidably leaving his fellow crewmember to crash, he could be proud enough of a mission accomplished to model his presidency on it; after murdering hundreds of thousands in Iraq and bailing out his friends' banks, he could again be proud, leaving the presidency to rejoin what he called the "real world." Those who are positioned, by various disjunctions from power, to see these contradictions do not labor out of some altruism or dedication to truth but because we are the ones left in the plane.

In times of crisis and great change the cost of various fictions becomes prohibitive, even for those who have traditionally been charged with maintaining them. It is not that Western Culture will be saved or lost (it will be both and neither; its identity has never been anything but a selective fiction); it is that laboring under notions of saving and losing—turf protection, damage control—has become more destructive, while the ongoing necessity of inventing more workable fictions has become more acute. Strategies which embrace contradiction will continue to be important: seeming to bite the hand that seems to feed us (whether an authorizing identity or discursive position), seeking to participate fully in a set of power relations from which our disjunction is also our enabling condition, and being driven rather than paralyzed by the double impossibilities of the detached ("ivory-tower") and the fully engaged ("organic") intellectual.

The human has been configured as a tribal circle gathered around the fire amid the looming darkness of a dangerous world, as the party of revelers sequestered from the plague, as the exclusive club of the Human, complete with all the rights and privileges pertaining thereunto (for example, the right to eat non-members of the club and the privilege not to be eaten). It is only partially our membership in the club that enables us to contest the rules, to beg to differ on how one must "assume the position" (take up the various crosses of identity, power, gender, authority). It is also because the darkness looms within the circle in a more virulent form, because some of the some that we are have been excluded; it is through multiple articulations among the constitutive roles of these others. Because otherness is not additive in the traditional sense, there is no "best" representative of the posthuman. Posthumans have been multiply colonized, interpenetrated, constructed—as well as paradoxically empowered—but neither virtue nor vice attaches automatically to this multiple position.

The posthuman does not necessitate the obsolescence of the human; it does not represent an evolution or devolution of the human. Rather it participates in re-distributions of difference and identity. The human functions to domesticate and hierarchize difference within the human (whether according to race, class, gender) and to absolutize difference between the human and the nonhuman. The posthuman does not reduce difference-from-others to difference-from-self, but rather emerges in the pattern of resonance and interference between the two. The additive other (who is subordinate in several systems at once) is not necessarily the geometrically other of the posthuman, who may well be "between between" in a single system. As a friend of ours likes to say, "I'm a feminist at a heavy metal concert and a metal advocate at a feminist meeting."

Family?

The human tribe can never again be family. Postfamilial bodies celebrate the end of His-and-Her matching theories that endlessly revolve around the miserable imagined unit, the imagined comm-unity of an imagined kinship in an imagined house with an imagined dog and two (if only) imagined children. Still, the story of the victory of the middle class and the hegemony of its family, discipline, and rationality as unmarked universals is as exaggerated as the story of their imminent demise. The shift in the balance of powers from the coercive to the disciplinary did not, of course, happen succinctly or uniformly, but unevenly and never completely. The rule of capitalism and the disciplinary power that has been its vice president is both total (no space is free of it) and partial (it does not reign uncontested in any of its locations). But its transnational and multi-dimensional ubiquity—its explosion to the horizon of global culture—can also be the precondition for other histories and powers to come into their own.

Lacan located the birth of human culture in the knowledge that heterosexual intercourse produces babies: the Name of the Father is secured by a system specifying who may be allowed to fuck what and how, producing mandates, prohibitions and selective freedoms in the circulation of fluids—breast milk, semen, money, gifts, information. The bio-taxonomy of species (and the order of knowledges of which it is an artifact) may be described as a similar set of mandates and prohibitions, along with the various "internal" divisions that it authorizes (species, order, family, genre, gender, divisions into sexual or asexual reproduction, warm and cold blood, etc.). Taxonomical discipline trains the branches of the genealogical "Tree of Life" to diverge neatly. Discursive bodies allow no such neat distinctions; they are both warm-blooded (self-regulating) and cold-blooded (sensitively dependent on their environments); both sexually and asexually reproduced. In any case, the ecology of interdependence problematizes the role of fucking in the life of species. When farting cows can be postulated as leading to catastrophic global climactic changes, who're you gonna call? A climatologist, a zoologist, a nutritionist, a Buddhist? What discipline has jurisdiction? If, magnified by technological interconnections, fear and hope can sweep across global stock markets as easily as they do across the Romantic humanist heart, shall we say humanism is dead or has reached its apotheosis?

If human reproduction, at least for the time being, necessarily involves the union of a sperm and egg, we are not created in, nor reducible to, their image (one per customer, please). Beyond the "little creatures . . . of love" of the Talking Heads song, allowed to name both sperm/eggs and adult bodies, is Dorion Sagan's "metametazoan," a multiple creature afloat in the non-complementary "omnisexuality" of bacterial exchanges, via which "the body becomes a sort of ornately elaborated mosaic of microbes in various states of symbiosis" and "health is less a matter of defending a unity than maintaining an ecology." Even so, the posthuman as "metametazoan" cannot therefore be subject to a "one-to-one linkage or reliably complete mapping" either with the multiplicities of microbes or with the planet Earth ("Gaia") conceived as a single/multiple organism (Sagan 369, 379). Posthumanities is alive to the ongoing danger of being shackled to the Great Chain of Being.

In practice as well as paradigm, sperm and egg unions have been repositioned. There are in any case multiple ways to bring about this union (tax incentives, in vitro fertilization, ideologies of family, turkey basters, etc.), or inhibit it (condoms, operations, tight jeans, abstinence, queer practices, etc.), and none of them are entirely reversible or irreversible. How can an Aristotelian hierarchization of causes separate the role of "the body" in reproduction from that of economy, technology, ideology, fashion? If biological reproduction is merely one possible function of one possible kind of fucking, as well as merely one of the many kinds of reproduction required to perpetuate the code of the human, then there is a curious lack of specificity in the term "fucking," a lack of coherence among its connotations, its variable association with pleasures and pains, with reproduction, with specific penetrations or frottages, with rhythmic frictions. What is allowed to be fucking? If the dissociation of female orgasm from generation that Laqueur locates in the late eighteenth century (1987) is what eventually allows female orgasm to signify the unspeakable ("jouissance") and unlocalizable mystery and the unreliability of signifiers, while male ejaculation (as in the "cum shot" of masculinist pornography) comes to guarantee the self-evidence of desire and truth in the binary of yes-or-no; this binary axiologization never could direct the traffics among power, pleasure, and bodies—traffics which include but are by no means exhausted by female ejaculation, sex-without-orgasm, orgasm-without-sex, sex-without-ejaculation, ejaculation-without-orgasm, reproduction-without-sex, sex-without-fucking, practices in which genitalia can become fetishes or second-order metaphors (a process impossible by definition in the one-way law of Freudian displacement and condensation), and so on. It becomes possible to assert a non-relation between fucking and reproduction—the relation upon which patriarchal humanity is predicated—partly because of the diversity of sexual practices, partly because of technological options, but mainly because the point where they converge is no longer an adequate anchoring point for a meaningful or workable system. Likewise, responsibility for conception and contraception, no less than for postnatal care, is not given but assigned.

The climacteric of the human dinosaur is a dangerous time, but no more than any other. The dying dinosaur still thrashes his tail, taking out hundreds of thousands in the process. Some of us cannot resist the risk that gnawing its scaly flesh entails; others strive to go about their business in discursive ecosystems in which the dinosaur could never compete, but all of us live in his shadow.

The infamous "family values" debate of the 1992 U.S. presidential election will be remembered as the discursive moment in which conservatives lost their hold on the imaginary place called "home." In what Jameson calls the homeopathy of postmodernism—the resistance through indulgence—family values dissolved at the touch. As soon as conservatives actually described the family they had in mind, its very visibility ruined its power as an ideological imaginary: there really is *no place* like "home." Discursive power operates from the imaginary, and identity registers its moment of failure. If the failure of "family values" has allowed a little sliding in what counts as "family," it has also bipartisanized the crusade on their behalf, making opposing positions still more difficult to articulate.

The posthuman repudiates the psychoanalytical and so the posthuman is also postpsychic, beyond any therapy that attempts to rectify the disorder and illogic of desires with health, purity and stability. Above all, purity dissolves in extrafamilial relations, where the body in culture is always a viral body, a time bomb of symptoms. Posthumanities embrace a radical impurity that includes the pure without privileging it. Extrafamilial desire exposes the family as a magic trick pulled by science and sustained by social science. Mommy and daddy are not sexy, and the Freudian family sitcom isn't funny anymore.

Aliens

If the human is dead, the alien, the other, goes with it. Or does it? What is different about the alien? Does posthumanity prop itself up against a human body or does it cannibalize the human?

David Cronenberg's films refuse to grant the category of human any particular primacy over other identities that jockey for position within the body. In *The Fly,* the scientist played by Jeff Goldblum revels in the disintegration of his human form, collects his human parts and creates a museum/mausoleum in his bathroom medicine cabinet. The human is emphasized here as a scientific showcase, a medical exhibit, a show of force but always a threatened constituency of body parts and reason. Goldblum becomes more and more repulsive, more and more likable and interesting as his form becomes fly. When he merges fly/human with the genetic structure of the computer and its attendant hardware, the triple other of animal/human/machine cannot slouch anywhere to be born but only abjectly crawl and beg to be killed; posthuman embodiment is frustrated seductively in the final instance in order to be nurtured in an imaginary or perverse reading the film can only insinuate. In any case, the human has been reduced to a moment, but not an evolutionary moment: it is a moment of flesh that interrupts a more intimate relation between body and machine.

In *Dead Ringers* the male subject is two male subjects who disintegrate because they find out that the inside of the body, specifically the inside of a woman's body, is mutant, beautiful, mesmerizing, infertile, and in/

human. Claire's infertility refracts the terrible fertility that produced the male subjects as not one baby but two. The twin gynecologists tremble before the gothicization of a body they know scientifically but not sexually. Again, the film's work is apparently negative; the self-sufficiency of male narcissism and the body that it codes is imploded through its oblique contact with its other, but the powerful identity-vacuum produced by this very thorough implosion into abjection is exactly where and how the film invites the posthuman to emerge. Recognition of a posthuman agenda requires new protocols for reading the positivity of horror and abjection, not as representational (as pedagogical object-lessons: don't try this at home) but as functional dysfunctions that make other things happen.

Catachresis

When Aristotle described "man" as a "featherless biped," Diogenes confronted him with a plucked chicken. To assert, in the spirit of this vaudeville philosophy, that humanity (and the human body) is a catachresis—a term unable either to ground itself adequately in a referent or to assert a common logic to unite its various referents—is a good first step, but the imaginary closure of the category of the human, even or especially if perpetually deferred, has very real functions. Unlike the human subject-to-be (Lacan's "l'hommelette"), who sees his own mirror image and fixed gender identity discrete and sovereign before him in a way that will forever exceed him, the posthuman becoming-subject vibrates across and among an assemblage of semi-autonomous collectivities it knows it can never either be coextensive with nor altogether separate from. The posthuman body is not driven, in the last instance, by a teleological desire for domination, death or stasis; or to become coherent and unitary; or even to explode into more disjointed multiplicities. Driven instead by the double impossibility and prerequisite to become other and to become itself, the posthuman body *intrigues* rather than desires; it is intrigued and intriguing just as it is queer: not as an identity but because it *queers*. Queering makes a postmodern politics out of the modernist aesthetics of "defamiliarization." "What intrigues me," k.d. lang asserts, "is being alternative and completely conformist at the same time" (98).

Queer

David Wojnarowicz, in *Close to the Knives: A Memoir of Disintegration,* writes:

> Realizing that I have nothing left to lose in my actions, I let my hands become weapons, my feet become weapons, every bone and muscle

> and fiber and ounce of blood become weapons, and I feel prepared for the rest of my life. (81)

The violence of a specifically queer posthumanity is realized when what Foucault calls the "reverse discourse" becomes something else, something more than the "homosexual talking on his/her own behalf." The reverse discourse ceases to be simply "the reverse" when it begins to challenge and disrupt the terms offered to it for self-definition. Coalition across what we have called the collectivity of someness creates a necessary space for queer articulations.

The AIDS body, for example, crumbles and disintegrates with the disease, but as Wojnarowicz shows, it also produces fear in those who do not have AIDS; it not only disintegrates, in other words, it produces disintegration at large. Disintegration as a political strategy attacks the oppressive imaginary gulf between the eternalized and "safe" body and the body at risk, the provisional body; it is this differential that constantly attempts to construct the Person-With-AIDS as "already dead," and beyond the human loop. Disintegration operates like a virus and infects people with fear of AIDS, exerting a weird kind of power, harnessed by ACT UP. The PWA, the junky, the homeless person, the queer in America also has power: as Wojnarowicz puts it, we have the power to "wake you up and welcome you to your bad dream." Queer tactics are not pacifist, embracing instead the "by any means necessary" approach: self defense and more. This is not simply an agenda of physical intimidation but a Foucauldian tactic of "discipline and punish," inspiring fear without actually laying a finger on anyone.

"Fear," Jenny Holzer writes, "Is the most Elegant weapon." *Close to the Knives* is really a manifesto for action, a proposal designed to strike fear into right-wing hearts; it is a call to arms, a call to live—to acknowledge that we live—close to the knives and close to the edge of violence. People who die of AIDS die violent deaths and Wojnarowicz proposes to make this violence visible.

The frame of reference within Wojnarowicz's personal holocaust is *viral:* the virus becomes an epistemology all its own, dividing the world into carriers and infected versus the possibly or potentially infected. The randomness of the disease means that everyone is affected by the infection of so many. This epistemology—knowing one's identity by measuring one's distance to or from the possibility of infection—opens up a window on other forms of knowing, on what he calls: "the unveiling of our order and disorder." Being Queer in America is a posthuman agenda.

At one point in Wojnarowicz's book, he describes videotaping the death of his friend in order to give the man a virtual existence beyond the grave.

Of course, Wojnarowicz's writing is also a technology that extends the body beyond death and beyond the disintegration of the body. Technologies that remake the body also permeate and mediate our relations to the "real": the real is literally unimaginable or only imaginable within a technological society: technology makes the body queer, fragments it, frames it, cuts it, transforms desire; the age of the image creates desire as a screen: the TV screen is analogous to self, a screen that projects and is projected onto but only gives the illusion of depth.

The image of an AIDS-related death being captured on film returns us all too quickly to U2's world of Zoo TV and its invitation to the reader to wonder which side of the lens she is on. While a connection between U2, an international mega-band, and Wojnarowicz, a queer artist dying of AIDS, may be arbitrary and coincidental, an odd image binds the two together. On the ZOO TV tour, U2 sold T-shirts featuring a silk-screened photo by David Wojnarowicz that appears as the cover of *Close to the Knives.* The photo shows buffalo stampeding over a cliff, and on the U2 T-shirt the Wojnarowicz caption, "Smell the flowers while you can," is scrawled underneath. The buffalo jumping to their doom, slipping off the edge of the earth and leaving their prairie zoo, resembles the medical zoo produced by the AIDS pandemic. This zoo cages AIDS-infected bodies and then drives them over the cliff. Smelling the flowers while you can means not simply hedonistic abandon but staving off apocalypse with pleasure. And then making your apocalypse one that requires witnesses.

"I'm carrying this rage like a blood-filled egg and there's a thin line between the inside and the outside a thin line between thought and action and that line is simply made up of blood and muscle and bone" (Wojnarowicz 161). Wojnarowicz trips over the line between inside and outside; he finds the meaning of his slow death in the anger that eats away at the human and the body and asks not for vengeance but for massive change and recognition that nothing is the same when you are dying a political death. The self disintegrates in this queer narrative into a posthuman rage for disorder and uncivil disobedience. For the queer narrator, rage is the difference between being and having: it is a call to arms, a desire that the human be roughly shoved into the next century and the next body and that we become posthuman without nostalgia and because we already are.

Quakes: The After Shock

Bodies depend on a network of signifying relationships. Following the San Francisco earthquake of 1990, there was a sharp rise in the battering of women by their husbands and boyfriends. The poor and homeless suffered disproportionately from the loss of their temporary shelters, often situated in old and substandard buildings, and from the diversion of social services. Nine months after the quake, area hospitals reported a sharp rise in the birth rate. In other words, the same people got fucked as usual, only more so. Far from being a "natural" event, the earthquake operated to confirm and reinforce the social distribution of violence. The discursive tremors in what had been considered the transhistorically stable ground of the body will not be so easily channeled.

Posthuman bodies never/always leave the womb. The dependence or interdependence of bodies on the material and discursive networks through which they operate means that the umbilical cords that supply us (without which we would die) are always multiple. The partial re-configurability of needs means that our navels are multiple as well. You can kill a significant portion of a country's inhabitants by disabling the country's "infrastructures" more economically than by shooting people; fertility treatments are less effective than tax incentives to produce babies; the Human Genome Project will do less to increase overall health than the redistribution of health care and wealth; changing how you walk and talk and dress and who and how you fuck changes your gender as well as surgery. These strategic assertions move the question from the dependence or contingency of bodies on the discursive networks in and by which they operate, to a refusal to distinguish absolutely or categorically between bodies and their material extensions.

Posthuman bodies were never in the womb. Bodies are determined and operated by systems whose reproduction is—sometimes partially but always irreducibly—asexual: capitalism, culture, professions, and institutions, and in fact sexuality itself. It is not merely that environmental factors are downloaded into the gene as the privileged mediator of bodily reproduction, but that the gene itself is everywhere. The localized and privileged gene promulgated by the Human Genome Project is a fetish because it hysterically displaces and condenses causality; hysterically because it serves to organize Big Science itself into the image of its fetish, an articulated control mechanism, each bit doing its part. If recent initiatives to locate the "origin" of violence in the "real" of the fetishized gene are matched, predictably, by equally laughable attempts to find the American violence gene in the "representational" space of television imagery; the diversionary repressive strategies that generate and are generated by these initiatives may not be so funny.

Against such initiatives, the current proliferation of books and articles on "the body" participate in a series of epistemic changes of which the body is both seismograph and epicenter. But the story that begins two hundred years ago with *The Birth of the Clinic* and *The Making of the*

Modern Body and ends, as we speak, with "The Death of the Author," *The Closing of the American Mind*, and *The End of History* is, after all, only the story of *a* body of discourse that always hysterically believed that it would die if its definite article were cut off, or revealed to have been detachable all along.

In *The Birth of the Clinic*, Foucault suggested that the late eighteenth-century shift in power/knowledge was succinctly enacted when doctors stopped asking their patients, "What is the matter with you?" and began to ask "Where does it hurt?" We add a third question: what is happening to your body?

Bodily masternarratives authorize a very narrow range of responses: that it is maturing or evolving or deteriorating or remaining the same, becoming dependent or independent; that it is threatened by, succumbing to or recovering from illness; that it is gaining or losing, for good or ill, various features or functions (weight, hair, muscles, mobility, etc.); that it is growing, reproducing, dying.

This range of authorized answers is *noise* for the purposes of our inquiry, and for most of what we feel is significant about what is happening to our bodies. What comes after the human is not another stage of evolution but a difference in kind. How is your body changing in kind? In small ways: I had my ear pierced (the topology of my body is changing; there's another hole all the way through it; my body is the earring of my earring). I got a tattoo (I participate in the cultural marking of my body). In other ways: it is changing its gender or its sexuality; that is, my sexual practices are re-configuring my body. I am becoming variously cyborgized (re-integrated with machine parts or across various networks). It is changing its dimensions, not by getting smaller or larger, but by being rhythmed across different sets of relations.

The transnationalization of culture has reached such a point that local traditions tend to be transformed (fossilized, commodified) into second-order phenomena: the bodies of our ancestors line the medium in which we now swim; the reef of culture is made of their skeletons. Those who resist the inroads of transnational capital and culture (in the name of national or ethnic integrity, appropriate technology, human-scale), and those who seek to make it habitable are not simply opposed, though articulations between them may be tendential; for example, those who find Mall Culture oppressively difference-leveling, and those who walk the Malls to recode and reconstitute them into a viable public sphere. Posthumanity is not about making an authentic culture or an organic community but about multiple viabilities.

When Marx imagined being able, in a postcapitalist utopia, to "fish in the morning, rear cattle in the afternoon and criticize in the evening, just as I wish, without ever becoming fisherman, farmer or critic" (160), he imagined a world in which the division of labor would neither divide people from themselves nor from each other; a world of practices without identities. To be able to (insert whatever sexual practice you wish) without becoming gay or straight, man or woman, requires not a productivist revolution that demands more options (more sexualities and genders, more discursive hybrids), but one which queries and queers the ways that the options are articulated and policed.

Queer, cyborg, metametazoan, hybrid, PWA; bodies-without-organs, bodies-in-process, virtual bodies: in unvisualizable amniotic indeterminacy, and unfazed by the hype of their always premature and redundant annunciation, posthuman bodies thrive in the mutual deformations of totem and taxonomy. We have rehearsed the claim that the posthuman condition is upon us and that lingering nostalgia for a modernist or humanist philosophy of self and other, human and alien, normal and queer is merely the echo of a discursive battle that has already taken place—and the tinny futurism that often answers such nostalgia is the echo of an echo. We stake our claim between these echoes and their answers.

Works Cited

Bhabha, Homi. "Signs Taken for Wonders," in *"Race," Writing, and Difference,* ed. Henry Louis Gates, Jr. (Chicago: University of Chicago Press, 1986).

Butler, Judith. "Imitation and Gender Insubordination," in *inside/out: Lesbian Theories, Gay Theories,* ed. Diana Fuss (New York: Routledge, 1991).

Deleuze, Gilles, and Felix Guattari. *A Thousand Plateaus: Capitalism and Schizophrenia,* trans. Brian Massumi (Minneapolis: University of Minnesota Press, 1987).

Derrida, Jacques. *Writing and Difference,* trans. Alan Bass (Chicago: University of Chicago Press, 1978).

Haraway, Donna. "Situated Knowledges" and "A Cyborg Manifesto," in *Simians, Cyborgs and Women* (New York: Routledge, 1991).

lang, k.d. Interview, in *Interview* magazine, 12/92.

Laqueur, Thomas. "Orgasm, Generation and the Politics of Reproductive Biology," in *The Making of the Modern Body,* ed. Catherine Gallagher and Thomas Laqueur (Berkeley: University of California Press, 1987).

Marx, Karl. "The German Ideology," in *The Marx-Engels Reader,* second ed., ed. Robert C. Tucker (New York: Norton, 1978).

Sagan, Dorion. "Metametazoa: Biology and Multiplicity," in *Incorporations* (New York: Zone, 1992).

Wojnarowicz, David. *Close to the Knives: A Memoir of Disintegration* (New York: Vintage, 1991).

Wordsworth, William. Preface to *Lyrical Ballads,* 1800, in *Poetical Works,* New Edition, ed. Thomas Hutchinson (Oxford: Oxford University Press, 1936).

PART I:

MULTIPLES

ONE

Identity in Oshkosh

Allucquere Rosanne Stone

The name is the end of discourse.

—Foucault

From the *San Francisco Chronicle:*

> On July 23, 1990, a 27-year-old woman filed a complaint in Oshkosh, Wisconsin charging that Mark Peterson, an acquaintance, raped her in her car. The woman had been previously diagnosed as having Multiple Personality Disorder (MPD). She claimed that Peterson raped her after deliberately drawing out one of her personalities, a naive young woman who he thought would be willing to have sex with him.

Cut to the municipal building complex in Oshkosh, Wisconsin. Outside the courthouse, gleaming white media vans line the street, nose to tail like a pod of refrigerators in rut. A forest of bristling antennae reaches skyward, and teenagers in brightly colored fast-food livery come and go bearing boxes and bags; the local pizza joints are doing a land-office business keeping the crews supplied. The sun is very bright, and we blink as we emerge from the shadows of the courthouse. "Jim Clifford would have loved this," I comment. "I wonder what the Mashpee courthouse looked like during the trial he was researching."

"Where's Mashpee?" my friend asks.

"In New England. The town of Mashpee was originally an Indian village. The Mashpee Indians deeded some land to the settlers, and the settlers eventually took over everything. A few years ago the surviving Mashpee families sued the town of Mashpee to get their land back, claiming that it had been taken from them illegally. When it finally came to trial, the government argued that the case revolved around the issue of whether the Mashpee now were the same Mashpee as the Mashpee then. In other words, were these Mashpee direct descendants of the original Mashpee in an uninterrupted progression.

"So the issue really being argued was, just what in hell is cultural continuity, anyway? Is it bloodline, like the government wanted it to be, or is it the transmission of shared symbols and values, like the view that the Mashpee themselves seemed to hold?

"That's why I find this trial so interesting, because what they're arguing here is both similar and different, and what's happening here both resonates and clashes with the Mashpee case in important ways."

While we stood in line there were a million-and-one other things I wanted to add. For example, the idea that personal identity is so refractory is a culturally specific one. Changing your name to signify an important change in your life was common in many North American cultures. Names themselves weren't codified as personal descriptors until the Domesday book. The idea behind taking a name appropriate to one's current circumstance was that identity is not static. Rather, the concept of one's public and private self, separately or together, changes with age and experience (as do the definitions of the categories public and private); and the name or the label on the identity package is an expression of that. The child is mother to the adult, but the adult is not merely the child a bit later in time.

Retaining the same name throughout life is part of an evolving strategy of producing particular kinds of subjects. In order to stabilize a name in such a way that it becomes a permanent descriptor, its function must either be split off from the self, or else the self must acquire a species of obduracy and permanence to match that of the name. In this manner a permanent name facilitates control; enhances interchangeability . . . if you can't have a *symbolic* identity (name) that coincides with your actual state at the time, then your institutionally maintained or *fiduciary* identity speaks you; you become the generic identity that the institutional descriptors allow.

Here in Oshkosh, instead of asking what is a culture, the unspoken question is what is a person. We all say "I'm not the person now that I was then," but as far as not only the government but everyone else is concerned, that's a figure of speech. In Mashpee exactly the opposite was being argued: whether the disparate lived experiences of individual members of a continually negotiated cultural system or an imagined cultural "unit" converged, through a legal apparatus transculturally imposed, on a unitary fiction, the fiduciary entity called the Mashpee tribe. In this trial, we have disparate experiences of individual social identities having at their focus a physical "unit," a fiduciary entity called the person, whose varying modes of existence both support and problematize the obduracy of individual identity and its refractoriness to deconstruction.

On this particular day, the first day of what by anybody's definition could be called the spectacle of multiplicity, everyone is getting their fifteen minutes' worth, their own little niche in the spectacle as multiplicity and violence get processed through the great engine of commodification just like everything else. Reporters from media all over the world are interviewing everything that moves. There are only so many people available in Oshkosh, and after exhausting whatever possibilities present themselves in the broad vicinity of the municipal complex, in a typical paparazzi feeding frenzy the media begin to devour each other. On the lawn not far from the courthouse doors Mark Blitstein, a reporter for the Oshkosh *Herald,* a small local newspaper, is grinning broadly. "I was just interviewed by the BBC," he says.

The cult of Isis reached full flower in Egypt at around 300 B.C.E., in the New Kingdom during the Persian Dynasties. The consul Lucius Cornelius Sulla brought the Isis myth to the Roman empire in 86 B.C.E., where it took root and flourished for nearly 700 years, becoming for a time one of the most popular branches of Roman mythology. The last Egyptian temples to Isis were closed sometime in 500–600 C.E.

The outlines of this familiar myth are simple: At first there existed only the ocean. On the surface of the ocean appeared an egg, from which Ra, the sun, was born. Ra gave birth to two sons, Shu and Geb, and two daughters, Tefnut and Nut. Geb and Nut had two sons, Set and Osiris, and two daughters, Isis and Nephthys. Osiris married his sister Isis and succeeded Ra as king of the earth. However, his brother Set hated him. Set killed Osiris, cut him into pieces, and scattered the fragments over the entire Nile valley. Isis gathered up the fragments, embalmed them, and resurrected Osiris as king of the netherworld, or the land of the dead. Isis and Osiris had a son, Horus, who defeated Set in battle and became king of the earth.

In his foundational work in abnormal psychology, *Multiple Personality Disorder,* Colin Ross makes the point that the Isis/Osiris myth illustrates the fragmentation, death, healing, and resurrection of the self in a new form. He probably chose this as his representative morphotype because it is more widely familiar than many other versions of what Joseph Campbell called the resurrection mythoid, an iconic fragment of human belief systems which Campbell asserted recurs across cultural boundaries. (Campbell had other problems with his theory that need not concern us here. We would be more likely to look to structuralism for similar theoretical approaches.)

Ross used the Osiris myth as a specific therapeutic model. He maintained that the MPD patient suffered from an Osiris complex, rather than an Oedipus complex. His abandonment of the Oedipus complex as a useful explanatory model stems from his reading of Freud's interpretation of the case of Anna O. and Freud's repudiation of the seduction theory following the publication of *Studies in Hysteria.* Ross's rationale is partly one of explanatory economy; he points out that the Oedipal model is what hackers would call a kluge—a complex, unwieldy, and aesthetically unsatisfactory patch that has the singular virtue of getting the job done—and that the Osiris model (not to mention the accompanying Isis model which would replace the Elektra complex) provides a much simpler and more elegant explanatory framework for multiple personality.

There are certainly enough varied opinions about what in hell is going on here to supply a very large number of theoreticians. The knots of professionals of various stripes engaged in muted or heated discussions call to mind the gedankexperiment of setting an infinite number of monkeys to the task of writing the complete works of Shakespeare. One of the psychologists observing the trial commented, "There's an awful excess of attention being paid to MPD these days. You know, in many ways it's being grossly overdiagnosed. And people are being channeled into it . . . it's like your most recent designer disease."

"Clinically speaking, does MPD have any positive aspects?"

"Well, it can be a way to get attention because of its fashionability in some therapeutic circles. There's no doubt that Sarah is a person who is not well. But she's learned to channel her illness so it gets attention. Or maybe she gets attention. But that way of dealing with a psychological problem has its own difficulties. It's also self-damaging. Part of her way of expressing it is to burn herself with cigarettes. Then her other personalities wonder how she got burned."

"Is there a possibility that she was acting? To get attention?"

He shook his head, looking thoughtful. "If she was acting, it was a hell of a brilliant job. And if she wasn't acting, then there was something else going on that was quite fascinating. Her vocabulary and demeanor, for instance . . . over time and place, they're consistent within a personality."

"How can you be sure that a particular person really has MPD and isn't faking it for some reason?"

"In many cases it's terribly hard to say . . . frequently difficult to make the call. Most MPDs are very intelligent. I'd think the more intelligent you were, the better you'd be able to fake something like that. If you were mentally ill anyway and knew it, there'd be excellent reasons to get a designer disease. You might be worried about getting lost in the state hospital system, and coming up with symptoms of MPD is a hell of a good way to get lots of attention quickly. If I were committed to a state facility, I'd try to generate a good case of MPD for myself just as fast as I could. That kind of thing can easily make the difference between life and death in some places, or between a reasonably comfortable life and being zombified by compulsory meds twenty-four hours a day."

"I couldn't say that I was absolutely sure just what MPD is, how it works, or really anything as simple as diagnostic procedures that worked in every case. A good part of what we're seeing here is a very tight interaction between the patients and the doctors, where a certain amount of the syndrome is occurring in the interactions between them, and that makes it very difficult to tell what's really going on. Do you get MPD when you're diagnosed or when you're two years old? I'd like to find out, in a definitive way, but it gets more difficult every day. The thing is taking on a life of its own."

"But in this case, at least, there's not much argument about whether the incident between Sarah (*who had no objection to her first name being used, but not her last*—A.R.S.) and Mark Peterson really happened . . . that's not what's at stake. After all, Sarah's condition wasn't exactly a secret. Her friends knew; the neighbors knew she had MPD. And Peterson . . . jesus." He shook his head. "It was clear to everybody that the guy was a real sleazebag, and that he was lying . . . after all, he bragged about it afterwards to friends. Hell, he bragged about it to the *cops*."

Rather than delegating the trial to a prosecutor, Winnebago County District Attorney Joseph Paulus is handling the case himself. He takes Petersen through the hoops, then doubles back. "Let's get back to your making love to Jennifer," Paulus says.

Peterson immediately corrected him. "I never said I made love to her."

Paulus looked faintly annoyed, went back to his table and riffled through a pile of papers there. "We have a statement here from you, in which . . . ah—" He found the page he was looking for—"you gave Officer Barnes a statement in which you claimed to have made love to Jennifer."

"Well, I . . . that's not right."

"Would you like me to read it to you?" Paulus was looking at the papers in his hand; he barely looked up when he said it.

"No. I mean I know what I said, and that statement was incorrect."

Paulus's eyebrows came up a trifle, but his expression didn't change. "In your statement to Officer Barnes you said that you and Sarah had discussed her multiple personalities before you went to the park, is that correct?"

"No, sir."

"You didn't talk about her multiple personalities with her?"

"No, sir."

"At any time?"

"That's right."

Paulus put most of the papers down, or rather slapped them down, and rounded on Peterson. He was plainly angry. "You allowed all kinds of untruths to go into this statement, didn't you?"

Peterson fumbled for a moment. "I was tired from a long day's work," he said.

Freud and Breuer published their classic work *Studies in Hysteria* in 1895. The book consisted of some case histories of their female patients and a number of chapters on theory. All the women described in the case histories had what would be theoretically described as dissociative disorders. In addition, most had been sexually abused. In Ross's view, Anna O., the subject of the most famous case history in the book, "clearly had MPD." Up to that point, Freud had considered these patients as suffering from the adult consequences of real childhood sexual abuse. His treatment took the reality of the trauma into account from both clinical and theoretical perspectives.

However, within a few years of publishing *Studies in Hysteria,* Freud repudiated the seduction theory which he had so carefully and effectively worked out. This point in Freud's development of his psychoanalytic theory has been a focus for study for some time; for example, Ernest Jones pointed out in his biography of Freud (1953) that many of the abusive fathers of Freud's dissociative female patients were part of Freud's social circle. This would have made it extremely awkward for Freud to state publicly that his patients had been sexually abused as children. Anna O.'s family lived in the Liechtensteinerstrasse, only one block from the Bergasse, where Freud both lived and worked. Breuer, for his part, was extremely uncomfortable with the sexual aspects of Anna O.'s symptomatology (Jones 247).

Edward Salzsieder, Peterson's attorney, started out with a novel and, until that moment, unthinkable idea. Salzsieder suggested that even though Wisconsin law forbade questioning a rape victim about her sexual history, such protection shouldn't extend to all of her other personalities. So he proposed questioning the other personalities—Franny and Ginger in particular—about *their* sexual histories. Many observers felt that this was one of the key points in the definition of multiple personality as a condition or state with legal standing other than as a pathology. For better or worse, Judge Hawley didn't think much of the idea. He did appreciate its complexity, though—"We're trying to split some very fine hairs here," he said—but he wasn't willing to take the idea so far as to impute autonomy to the multiples. "I do find," he said, "that the rape shield law applies to [Sarah] and all her personalities combined."

That threw Salzsieder back on his own resources. Deprived of the opportunity to question the personalities about their individual sexual exploits, he fell back on the strategy of attacking their legitimacy. To bring this off he needed to assemble a cadre of MPD infidels, unbelievers with legal and professional stature who, he hoped, could cast doubt on the whole idea of MPD. As it turns out, it wasn't difficult to do. All kinds of people were willing to testify on all sides of the issue. But Salzsieder was looking for a special person, someone who not only didn't believe in MPD but who could convince a court that MPD was a convenient fantasy, something that Sarah had read about and then adopted to excuse her promiscuous behavior. Eventually he came up with Donald Travers. Travers is from Wisconsin, a slightly balding man of medium build who when on the stand projects the proper blend of sober professionalism and easy believability that Salzsieder needed. Travers is an impressive infidel. He is an articulate speaker who is convinced that MPD is a medical hoax and whom Salzsieder had gotten to review Sarah's psychiatric records for the previous year.

Salzsieder started by getting Travers to attack the credibility of MPD as a diagnostic category. After Travers was sworn in, Salzsieder asked "How many psychologists actually have patients with MPD?"

"There's a band of very intense believers who have all the sightings, where the rest of us never see any," Travers said. "What I call the UFOs of psychiatry."

"In your professional opinion, what would you call Sarah's condition?"

Travers put his fingertips together like a character from a Perry Mason episode. "I would say that . . . I believe Sarah does have psychiatric problems, but her problems don't appear grave enough to fit within the DSM3 guidelines."

"She's well enough to know what she's doing. Is that what you mean? Responsible for her own actions?"

"That's correct."

In one of the foundational accounts of MPD, Colin Ross identifies the fragmentation of self and the transformation of identity that occurs across ethnic and cultural boundaries, all of which he lumps together under the rubric of "aberration." While his identification of this characteristic of human cultures is correct, his use of the rubric is peculiarly situated. Ross is interested in making a strong case for legitimizing MPD as a recognized medical phenomenon, and in so doing he seems to feel that he must explain away the problem of why many of the cultures he mentions in passing do not themselves pathologize MPD. Ross can perhaps be excused for pathologizing MPD *tout court*, because he evinces a genuine interest in assisting the individuals he has observed whose accommodation to buried trauma causes, in his words, more suffering than it prevents. I am primarily concerned here with how the phenomenon of multiple personality fits into a broader framework of cultural developments in which the abstract machine of multiplicity (in Deleuze and Guattari's words) is grinding finer and finer. Among the phenomena at the close of the mechanical age which are useful to note is the pervasive burgeoning of the ontic and epistemic qualities of multiplicity in all their forms.

It is the moment everyone in the courtroom has been waiting for. People had been standing in line since before dawn to assure themselves of seats in the courtroom. A few had brought folding chairs to use while they waited in the predawn chill. Some sat on beach blankets with thermos jugs of steaming coffee. The composition of the crowd was extraordinarily diverse.

After an agonizing wait while people chatted to each other with the same lively animation I associated with waiting for the start of a long-anticipated film, the bailiff called the room to order. The silence was instantaneous. "All rise," the bailiff called, and Hawley strode in, followed by the court stenographer.

Hawley sat down in the high-backed leather chair, squared something on his desk, and looked down from the bench at the packed courtroom, his glasses catching the light. The sound of people getting seated died away, and a hush again fell over the room.

For the most part Hawley had not said very much beyond what was required of him as presiding magistrate, but this morning he cleared his throat and made a brief introductory speech. His voice carried well in the room. It was a calm voice, not too inflected.

"Before we proceed any further, I want to make sure all the video and film equipment in this room is turned off and that all the cameras are down out of sight." He scanned the room slowly, more for effect than for surveillance, then continued in the same calm voice. "There has been an unusual amount of attention surrounding this case. The issues we are considering are of an unusual nature. But I want to make it clear to everyone here that this is not a circus. This is a very sensitive case. There may be some bizarre behavior that you have not witnessed before. But nothing should get in the way of this being a court of law, first and foremost. I know that I can expect you to behave appropriately."

Nods from the spectators. People settled deeper into their seats. The unusually large population of professionals among the spectators now made itself known as people reached into bags and briefcases for their yellow notepads, making the room bloom like a gray field dotted with buttercups.

Hawley nodded to Paulus. The silence deepened, if that were possible, and Paulus called his first witness of the day.

Sarah walked briskly to the stand. She seated herself and was sworn in. She put her hands in her lap and looked calmly at Paulus. This is the main event, I thought. It is what this whole thing is about, really. It is not columns in a newspaper. It is not theory or discussion. It is not soundbyte media hype. It is a young, calm, slightly Asian-looking woman in a white cotton sweater and a pale blue skirt.

Paulus stood a few feet in front of her, holding his body relaxed and still. He spoke to her in a normal conversational tone, not very loud but clearly audible in the silent room.

"Sarah, you've heard some testimony here about some events that took place recently in Shiner Park. Do you recall that testimony?"

Sarah nodded slightly, then added, "I do."

"Do you have any personal knowledge as to the events in the park?"

"No," Sarah said, "I do not." Her voice was quiet, flat, matter-of-fact.

"Who would be in the best position to talk about the events in the park that night?"

"Franny," Sarah said.

"Would it be possible for us to, uh—" Paulus hesitated and looked like he wanted to clear his throat, but he settled for an instant's pause instead and then continued—"meet Franny, and talk to her?"

"Yes," Sarah said, looking calmly at him. A beat or two. "Now?"

"Yes," Paulus said. "Take your time."

The silence was absolute. Faintly, from somewhere outside in the hallway, something metallic dropped to the floor and rolled.

Sarah closed her eyes and slowly lowered her head until her chin was resting on her chest. She sat that way, her body still, breathing slowly and shallowly. It seemed as though everyone in the room held a collective breath. The muted hush of the air conditioning came slowly up from the background as if someone had turned up a volume control.

Maybe five seconds passed, maybe ten. It felt like hours. Then she raised her head, and slowly opened her eyes.

She looked at Paulus, and suddenly her face was animated, alive and mobile in a way that it hadn't been a moment ago. The muscles around her mouth and eyes seemed to work differently, to be somehow more robust. She looked him up and down, taking him in with obvious appreciation. "Hel-*lo,*" she said.

"Franny?" Paulus said, inquisitively.

"Good morning," Franny said. She looked around at the windowless courtroom. "Or good afternoon—which is it?" Her phrasing was more musical than it had been, with an odd lilt to the words. It, too, was animated, but it didn't sound quite like an animated voice should have sounded. Also, on closer inspection it appeared that the more animated look of her features hadn't made it down into her body. Her posture, the way she held herself, the positions of her shoulders and legs and the relative tension in the muscles of her body, hadn't changed very much from Sarah's posture.

Paulus looked as if he wanted to feel relieved, but again he hid it quickly. "It's, uh, morning, actually," he said, in a conversational tone. "How are you today?"

"I'm fine. How are you?" The same lilt to the words.

"Just fine. Now I was just talking to Sarah a few moments ago, and I'd like to talk to you about what happened June ninth of 1990." He glanced up at Hawley. "But before we do that, the judge has to talk to you."

Hawley looked down at Franny. When she faced forward most of what he could see of her was the top of her head, but she turned now to face him. Her expression was hard to catch, but Hawley looked perfectly placid, as if swearing in several people in one body were something he did every day. "Franny," he said, "I'd like you to raise your right hand for me, please."

Hawley swore her in, his face impassive. It sounded like any other court ritual. When they got to the "so help you God" part, Franny said "Yes," they both lowered their hands, and she turned back to Paulus.

"What did he say?"

"He said it felt good. And I knew what I was supposed to do when he said that. I seen it on TV. People wiggling like that. And when a person says it feels good, the other person is supposed to say it feels good. So I put my arms around his back, and I said, "That feels nice."

"Did it feel nice?" Paulus asked.

"No," she said, sounding perplexed. "But you're supposed to say that, aren't you? It was on TV."

An important aspect of Freud's personal genius, and one with lasting import for the developing field of psychoanalysis, was his ability to construct a clinically plausible *and socially acceptable* theory that explained the phenomenon of adult dissociation and simultaneously denied the reality of childhood sexual abuse. This, which Ross refers to as Freud's unfortunate "metapsychological digression," was the theory of the Oedipus and Elektra complexes. Returning resolutely to his point, Ross asserts that the Osiris complex more clearly describes what happens in the etiology of MPD than the Oedipus or Elektra complexes can, and it does it with a minimum of description; if economy of representation counts for anything, the Osiris complex wins hands down. The trouble with the Osiris myth in a modern clinical frame, as Ross comments, is that "in our culture, the original agent of the fragmentation of self does not always receive divine retribution. . . . "

From the point at which Freud repudiated the seduction theory, and continuing forward almost to the present, psychoanalysts of Freudian persuasion considered patients with dissociative disorders of traumatic origin to be suffering from unresolved unconscious incestuous fantasies. This state of affairs has been treated at length by feminist scholars (e.g., Rivera 1987, 1988; Sprengnether 1985, among others).

We are still looking at traumatically produced MPD here, still using the final D to indicate that the thing is a disorder and nothing more. Just what is it, then, that we are looking at, and why is MPD so important to an examination of communication technology? More to the point, is there any room for non-traumatic multiplicity in any of these clinical accounts? At one point Ross, for example, answers this question almost dismissively and with complete self-confidence: "The term (multiple personality) suggests that it is necessary to debate whether one person can really have more than one personality, or, put more extremely, whether there can really be more than one person in a single body. Of course there can't . . . " (41). And here Ross misses some of the most crucial implications of his study.

Multiple personality (without the stigmatizing final D) is a mode that resonates throughout the accounts I present here. Ross's research both affirms and denies that mode in a complex way. He has a clear investment in affirming the reality of a clinical definition of multiplicity, and his views concerning Freud's problems in coming to grips with the probable etiology of clinical multiplicity (with the D) are useful in studying the influence Freud has had on the field of psychoanalysis. For reasons that I find not entirely clear, he dismisses out of hand the idea that there can be more than one person in a single body. At that point he appears to fall back on received social and cultural norms concerning the meaning of "person" and "body." Like the surgeons at the Stanford Gender Dysphoria Project, of whom I have written elsewhere, Ross still acts as a gatekeeper for meaning within a larger cultural frame, and in so doing his stakes and investments become clearer.

In this context, the context of multiplicity and psychology, it is useful to consider the work of Sherry Turkle. In her study *Constructions and Reconstructions of the Self in Virtual Reality*, presented at the Third International Conference on Cyberspace, Turkle notes:

> The power of the (virtual) medium as a material for the projection of aspects of both conscious and unconscious aspects of the self suggests an analogy between multiple-user domains (MUDs) and psychotherapeutic milieus. . . . MUDs are a context for constructions and reconstructions of identity; they are also a context for reflecting on old notions of identity itself. Through contemporary psychoanalytic theory which stresses the decentered subject and through the fragmented selves presented by patients (and most dramatically the increasing numbers of patients who present with multiple personality) psychology confronts the ways in which any unitary notion of identity is problematic and illusory. What is the self when it functions as a society? What is the self when it divides its labor among its constituent "alters" or "avatars"? Those burdened by posttraumatic dissociative syndrome (MPD) suffer the question; inhabitants of MUDs play with it.

In Turkle's context, the context of virtual systems, the question that Ross dismisses as, to him, obviously false—namely, can multiple selves inhabit a single body—is irrelevant. Compared to "real" space, in virtual space the socioepistemic structures by means of which the meanings of the terms "self" and "body" are produced operate differently. Turkle seizes upon this and turns it into a psychotherapeutic tool. Moreover, Turkle shows how the uses of virtual space as an adjunct to therapy translate across domains, beyond the virtual worlds and into the biological. What in this context might be called the ultimate experiment—plugging a person with MPD into the MUDs—has yet to be performed. Thus we have not yet observed one of its possibly hopeful outcomes: healing trauma, but preserving multiplicity; or perhaps more pertinent, creating discursive space for a possibly transformative legitimization of some forms of multiplicity. The answers to the questions posed above—why is MPD so important to an examination of communication technology, and is there room for non-traumatic multiplicity in clinical accounts—in fine are bound up with the prosthetic character of virtuality. The technosocial space of virtual interaction, with its irruptive ludic quality, its potential for experimentation and emergence, can be a problematic and hopeful domain of non-traumatic multiplicity. Turkle and others, myself included, are waiting to observe how the dialogue between non-traumatic multiplicity and clinical accounts emerges in a new therapeutic context.

"Do you remember what you talked about while he was there?"

"Most of it was small talk. I recall telling him that we were many, there were many of us in the body. I said we were multiple, that we shared the body. I told him about some of the others."

"How did he react to your telling him about some of the others?"

"He didn't seem surprised."

Although everyone had their own reason for being there, nobody could quite explain what their fascination with the courtroom scene was. I tried to make a few mental models, and it didn't work. The courtroom audience's behavior, though, was its own giveaway. Its attention was on the moment of rupture, conjoining the sacred and the forbidden. I am certainly not the first to label this moment the moment of interruption, when the seamless surface of reality is ripped aside to reveal the nuts and bolts by which the structure is maintained. Sarah was a liminal creature, marked as representing something deeply desired and deeply feared. In the same court an ax murderer would attract a certain ghoulish attention, but nothing like the fascination we were seeing here. On the principle that where one finds a circumstance which is a focus of the most intense emotional energy coupled with the least understanding of why it is such a focus, there is the place to dig, then it seemed clear enough that the moment Franny appeared was that moment.

Multiple personality, as it is commonly represented, is the site of a massive exercise of power and its aftermath, the site of a marshaling of physical proof that identity—of whatever form—arises in crisis. It vividly demonstrates the connection between the violence of splitting off a string of identities to the violence of representation under the sign of the patristic Word in a court of law. In order for the prosecution's strategy to work, the victim must manifest a collection of identities, each one of which is recognizable to the jury as a legal subject. We are witnesses to an exercise of power, to an effort to fix in position a particular subjectivity. Having thus been drawn to the grotesque—in this case, to the spectacle of the maimed persona—we might reflect on how we got here and where we were going when our attention was arrested.

First is the spectacle of violence at the margins, at the origins of subject construction. To make the discredited move from the local to the universal, in the violence by which the multiple subject is constituted in the medical syndrome we recognize the elements by which national identities have traditionally arisen—the consolidation of a sense of conscious autonomy in an act of violence, temporally and physically at the site of its application. We are witnesses to a spectacle that as civilized beings we would prefer not to acknowledge—a site at which the apparatus of production of subjectivity is laid bare—and the safe course is to view such a site as an aberration, as pathology, engendered by an unfortunate encounter with a sick author(ity). We fail to make the identification when confronted with a particular narrative of passage, that of recognizing the protagonist as oneself. We miss the lesson of how we came to be capable of being constructed as witnesses *ab origine,* miss comprehending our own violent origin.

The trial ends with Peterson's conviction.

This outcome is a mixed grill for the various interests surrounding the trial. While there are several points on which new law might have been written, two in particular are interesting in connection with the Peterson trial. One concerns the conflation of multiple personality with mental illness. Another concerns the legal status of each member of a multiple personality. Both of these relate to issues of how cultural meaning is constructed in relation to bodies and selves.

Ruth Reeves, Sarah's downstairs neighbor, is a woman with no particular investment in much of the debate. "I've met most of them (the personalities), and they're real," she says. "It's no different, really, than talking to a roomful of people."

"Do you think she's sick . . . mentally ill? I mean, multiple personality as a disease . . . "

"Well, her personalities mostly just seem to live their lives. It's not like one of them's a murderer or goes around busting up the furniture. Some of them aren't healthy for her, though. I hope therapy can help her, so she doesn't have to do things like eat crayons or burn herself. But—" She looked thoughtful for a moment, searching for words.

"You know," she said, "if the therapy turns out to be effective I'm going to miss the personalities. They're a wonderful bunch of folks."

The verdict upholds existing Wisconsin law. The law states that it is a crime to have sex with a mentally ill person if the person is so severely impaired that he or she cannot appreciate the consequences of their behavior, and if the other person knows of the illness. Because the trial made no attempt to separate the issue of MPD from issues of mental illness, the verdict reinforces the general conflation of multiple personality with mental illness. This seems natural to the great majority of mental health professionals who viewed the trial. A few, who perceived the opportunity to "decriminalize" MPD, are disappointed.

"Multiple personality" covers a broad range of phenomena, which includes within its spectrum such things as spirit possession. "Multiple personality disorder" is the official term for a condition which includes, among other things, blackouts. That is, only one personality is out at a time, and if there is a dominant personality it suffers memory gaps during the time the other personalities are out. "You find clothes in your closets that you have no memory of having bought, and worse yet, they aren't the cut or color you would ever think of buying," one multiple says. "You get court summonses about traffic violations you didn't commit, you wake up in the morning and find you have burns and bruises and you have no idea how or where you got them." In general the dominant personality is frightened and troubled by these occurrences. The dominant personality may also have difficulty coping in the world, and it is this maladjustment, or the fear and disorientation caused by the blackouts, that generally brings the person into the doctor's office.

At the other end of the spectrum are persons who also consider themselves multiples, but who do not suffer blackouts and who claim to retain awareness of what the alter personalities are doing when they are out. These persons find themselves in a difficult situation. If they assert their multiplicity, they fear being pathologized, so they tend to live "in the closet," like other marginalized groups. They live largely clandestine existences, holding regular day jobs and occasionally socializing with other multiples of similar type. They worry about being discovered and being forced to quit their jobs, or about being declared disabled or mentally incompetent. They have no common literature which unites them; the multiple equivalent of *The Well of Loneliness* has yet to be written. Their accustomed mode of existence, sharing a single body with several quasi-independent personalities, is emblematic of a fair percentage of everyday life at the close of the mechanical age.

TWO

Two Lessons from Burroughs

Steven Shaviro

Seattle, 1993. Don't believe the hype. I find myself stranded in this obsessively health-minded, puritanical, routinized, and relentlessly cheerful city, lifelines cut, lost without my vital supply of counteracting stimulants. Yes, some of the bands are still great, despite the insidious pressures of fame: Nirvana, Mudhoney, Seven-Year Bitch. But otherwise, nothing. I strain to hear the echo of Burroughs's silent scream: "What scared you all into time? Into body? Into shit? I will tell you: *the word*." But does anyone even remember? These prefabricated combinations of words, and these carefully crafted, HWP bodies, are all I can find, perhaps all there is. Organicism is a myth. Our bodies are never ourselves, our words and texts are never really our own. They aren't "us," but the forces which crush us, the norms to which we have been subjected. It's a relief to realize that culture is after all empty, that its imposing edifices are sound stage facades, that bodies are extremely plastic, that facial expressions are masks, that words in fact have nothing to express. Bodies and words are nothing but exchange-value: commodities or money. All we can do is appropriate them, distort them, turn them against themselves. All we can do is borrow them and waste them: spend what we haven't earned and don't even possess. Such is my definition of postmodern culture, but it's also Citibank's definition of a healthy economy, Jacques Lacan's definition of love, and J. G. Ballard's vision of life in the postindustrial ruins. So don't be a good citizen. Don't produce, expend. Be a parasite. Live off your Visa card, or scavenge in the debris.

With all this in mind, I want to propose a biological approach to postmodernism. Ethology rather than ethnology. As we know from Foucault, from François Jacob, and from Donna Haraway, "biology" as we understand it today is a very recent invention. But of course it works both ways. Every mutation in culture is a new state of the body. Technological changes, as McLuhan said, are alterations in the very nature of our senses and of our nervous systems. The inventions that make, say, genetic engineering practicable are themselves biological innovations. The conditions of possibility for postmodernism first evolved something like one million years ago, with the appearance in our hominid ancestors of what might be called the Ronald Reagan gene or meme: the program for deceiving others more effectively by at the same time deluding yourself. This allows you to project a powerful aura of absolute sincerity. Pull the wool over your own eyes, as the Church of the SubGenius puts it. But the Reagan strategy is only one move in a long history of manipulations, power grabs, and scams. Freud and Lacan to the contrary, there's nothing less "essentialist," less "organicist," more political, and more historically variable than our "anatomy" or "biology." I leave open for the moment the question of just how far this pronominal "our" extends.

Nobody understands these issues better than William Burroughs. All his major novels, from *Naked Lunch* (1959) to *The Western Lands* (1987), have explored the landscape of postmodern biology, with its deliriums and its terrors. That's why I invoke him as my guide in what follows. These "lessons" about language and about insects are only two of many to be learned from Burroughs. But a word of caution is in order. As we read in *Nova Express:* "And what does my program of total austerity and total resistance offer *you*? I offer you nothing. I am not a politician. . . . To speak is to lie—To live is to collaborate—There are degrees of lying collaboration and cowardice—It is precisely a question of *regulation*. . . . "

1. Language Is a Virus

"Which came first, the intestine or the tapeworm?" In this epigram, William Burroughs suggests that parasitism—corruption, plagiarism, surplus appropriation—is in fact conterminous with life itself. The tapeworm doesn't simply happen to attach itself to an intestine that was getting along perfectly well without it. Say rather that the intestine evolved in the way that it did just in order to provide the tapeworm with a comfortable or profitable milieu, an environment in which it might thrive. My intestines are on as intimate terms with their tapeworms as they are with my mouth, my asshole, and my other organs; the relationship is as "intrinsic" and "organic" in the one case as it is in the other. Just like the tapeworm, I live off the surplus-value extracted from what passes through my stomach and intestines. Who's the parasite, then, and who's the host? The internal organs are parasitic upon one another; the organism as a whole is parasitic upon the world. My "innards" are really a hole going straight through my body; their contents—shit and tapeworm—remain forever outside of and apart from me, even as they exist at my very center. The tapeworm is more "me" than I am myself. My shit is my inner essence; yet I cannot assimilate it to myself, but find myself always compelled to give it away. (Hence Freud's equation of feces with money and gifts; and Artaud's sense of being robbed of his body and selfhood every time he took a shit.) Interiority means intrusion and colonization. Self-identity is ultimately a symptom of parasitic invasion, the expression within me of forces originating from outside.

And so it is with language. In Burroughs's famous dictum, language is a virus from outer space. Language is to the brain (and to the speaking mouth and the writing or typing hand) as the tapeworm is to the intestines. Or even more so: it may just be possible to find a digestive space free from parasitic infection (though this is extremely unlikely), but we will never find an uncontaminated mental space. Strands of alien DNA unfurl themselves in our brains, even as tapeworms unfurl themselves in our guts. Burroughs suggests that not just language, but "the whole quality of human consciousness, as expressed in male and female, is basically a virus mechanism." This is not to claim, in the manner of Saussure and certain foolish poststructuralists, that all thought is linguistic, or that social reality is constituted solely through language. It is rather to deprivilege language—and thus to take apart the customary opposition between language and immediate intuition—by pointing out that nonlinguistic modes of thought (which obviously exist) are themselves also constituted by parasitic infiltration. Visual apprehension and the internal time sense, to take just two examples, are both radically nonlinguistic; but they too, in their own ways, are theaters of power and of surplus-value extraction. Light sears my eyeballs, leaves its traces violently incised on my retinas. Duration imposes its ungraspable rhythms, emptying me of my own thought. Viruses and parasitic worms are at work everywhere, multiple "outsides" colonizing our "insides." There is no refuge of pure interiority, not even before language. Whoever we are, and wherever and however we search, "we are all tainted with viral origins."

Burroughs's formulation is of course deliberately paradoxical, since viruses are never originary beings. They aren't self-sufficient, or even fully alive; they always need to commandeer the cells of an already-existing host in order to reproduce. A virus is nothing but DNA or RNA encased in a protective sheath; that is to say, it is a message—encoded in nucleic acid—whose only content is an order to repeat itself. When a living cell is invaded by a virus, it is compelled to obey this order. Here the medium really is the message: for the virus doesn't enunciate any command, so much as the virus *is* itself the command. It is a machine for reproduction, but without any external or referential content to be reproduced. A virus is thus a simulacrum: a copy for which there is no original, emptily duplicating itself to infinity. It doesn't represent anything, and it doesn't have to refer back to any standard measure or first instance, because it already contains all the information—and only the information—needed for its own further replication. Marx's famous description of capital applies perfectly to viruses: "dead labour which, vampire-like, lives only by sucking living labour, and lives the more, the more labour it sucks."

Reproduction (sexual or otherwise) is often sentimentally considered to be the basic activity and fundamental characteristic of life. But it is arguably more a viral than a vital process. Reproduction is so far from being straightforwardly "organic," that it necessarily involves vampirism, parasitism, and cancerous simulation. We are all tainted with viral origins, because life itself is commanded and impelled by something alien to life. The life possessed by a cell, and all the more so by a multicellular organism, is finally only its ability to carry out the orders transmitted to it by DNA and RNA. It scarcely matters whether these orders originate from a virus, or from what we conceive as the cell's own nucleus. For this distinction is only a matter of practical convenience. It is impossible actually to isolate the organism in a state before it has been infiltrated by viruses, or altered by mutations; we cannot separate out the different segments of DNA, and determine which are intrinsic to the organism and which are foreign. Our cells' own DNA is perhaps best regarded as a viral intruder that has so successfully and over so long a stretch of time managed to insinuate itself within us, that we have forgotten its alien origin. Our genes' "purposes" are not ours. As Richard Dawkins puts it, our bodies and minds are "survival machines" programmed for replicating genes, "gigantic lumbering robots" created for the sole purpose of transmitting DNA. Burroughs describes language (or sexuality, or any form of consciousness) as "the *human virus*." All our mechanisms of reproduction follow the viral logic according to which life produces death, and death in turn lives off life. And so remember this the next time you gush over a cute infant. "Cry of newborn baby gurgles into death rattle and the crystal skull," Burroughs writes, "THAT IS WHAT YOU GET FOR FUCKING."

Language is one of these mechanisms of reproduction. Its purpose is not to indicate or communicate any particular content, but merely to perpetuate and replicate itself. The problem with most versions of communications theory is that they ignore this function, and naively present language as a means of transmitting information. Yet language, like a virus or like capital, is in itself entirely vacuous: its supposed content is only a contingent means (the host cell or the particular commodity form) that it parasitically appropriates for the end of self-valorization and self-proliferation. Apart from the medium, there's no other message. But if language cannot be apprehended in terms of informational content, still less can it be understood on the basis of its form or structure, in the manner of Saussure, Chomsky, and their followers. These theorists make an equivalent, but symmetrically opposite, error to that of communications theory. They substitute inner coherence for outer correspondence, differential articulation for communicative redundancy, and self-reference for external reference; but by isolating language's self-relational structure or transformational logic, they continue to neglect the concrete and pragmatic effects of its violent replicating force. Both communicational and structural approaches try to define what language *is*, instead of looking at what it *does*. They both fail to come to grips with what J. L. Austin calls the *performative* aspect of linguistic utterance: the sense in which speaking and writing are *actions*, ways of doing something, and not merely ways of (con)stating or referring to something. (Of course, stating and referring are in the last analysis themselves actions.) Language does not represent the world: it intervenes in the world, invades the world, appropriates the world. The supposed postmodern "disappearance of the referent" in fact testifies to the success of this invasion. It's not that language doesn't refer to anything real, but—to the contrary—that language itself has become increasingly real. Far from referring only to itself, language is powerfully intertwined with all the other aspects of contemporary social reality. It is a virus that has all too fully incorporated itself into the everyday life of its hosts.

A virus has no morals, as Rosa von Praunheim puts it, talking about HIV; and similarly the language virus has no meanings. Even saying that language is performative doesn't go far enough; for it leaves aside the further question of what sort of act is being performed, and just *who* is performing it. It is not "I" who speaks, but the virus inside me. And this virus/speech is not a freestanding action, but a motivated and directed one: a command. Morse Peckham, Deleuze and Guattari, and Wittgenstein all suggest that language is less performative than it is imperative or prescriptive: to speak is to give orders. To understand language and speech is then to acknowledge these orders: to obey them or resist them, but to react to them in some way. An alien force has taken hold of me, and I cannot *not* respond. Our bodies similarly respond with *symptoms* to infection, or to the orders of viral DNA and RNA. As Burroughs reminds us: "the symptoms of a virus are the attempts of the body to deal with the virus attack. By their symptoms you shall know them. . . . If a virus produces no symptoms, then we have no way of knowing that it exists." And so with all linguistic utterances: I interpret a statement by reacting to it, which is to say by generating a symptom. Voices continually call and respond, invoke and provoke other voices. Speaking is thus in Foucault's sense an exercise of power: "it incites, it induces, it seduces, it makes easier or more difficult; in the extreme it constrains or forbids absolutely; it is nevertheless always a way of acting upon an acting subject or acting subjects by virtue of their acting or being capable of acting. A set of actions upon other actions." Usually we obey orders that have been given us, viscerally and unreflectively; but even if we self-consciously refuse them, we are still operating under their constraint, or according to their dictation. Yet since an order is itself an action, and the only response to an action is another action, what Wittgenstein ironically calls the "gulf between an order and its execution" always remains. I can reply to a performance only with another performance; it's impossible to step outside the series of actions, to break the chain and isolate once and for all the "true" meaning of an utterance. The material force of the utterance compels me to respond, but no hermeneutics can guarantee or legislate the precise nature of my response. The only workable way to define "meaning" is therefore to say, with Peckham, that it is radically arbitrary, since "*any response to an utterance is a meaning of that utterance.*" Any response whatsoever. This accounts both for the fascistic, imperative nature of language, and for its infinite susceptibility to perversion and deviation. Strands of DNA replicate themselves *ad infinitum.* But in the course of these mindless repetitions, unexpected reactions spontaneously arise, alien viruses insinuate themselves into the DNA sequence, and radiation produces random mutations. It's much like what happens in the children's game "Telephone": even when a sentence is repeated as exactly as possible, it tends to change radically over the course of time.

We all have parasites inhabiting our bodies; even as we are ourselves parasites feeding on larger structures. Call this a formula for demonic or vampiric possession. The great modernist project was to let the Being of Language shine forth, or some such grandiose notion. If the "I" was not the speaker, the modernists believed, this was because language itself spoke to me and through me. Heidegger is well aware that language consists in giving orders, but he odiously idealizes the whole process of command and obedience. We postmodernists know better. We must say, contrary to Heidegger and Lacan, that language *never* "speaks itself as language": it's always some particular parasite, with its own interests and perspective, that's issuing the orders and collecting the profits. What distinguishes a virus or parasite is precisely that it has no proper relation to Being. It only inhabits somebody else's dwelling. Every discourse is an unwelcome guest that sponges off me, without paying its share of the rent. My body and home are always infested—whether by tapeworms and cockroaches, or by Martians and poltergeists. Language isn't the House of Being, but a fairground filled with hucksters and con artists. Think of Melville's Confidence Man; or Burroughs's innumerable petty operators, all pulling their scams. Michel Serres, in *The Parasite*, traces endless chains of appropriation and transfer, subtending all forms of communication. (He plays on the fact that in French the word *parasite* has the additional connotation of *static*, the noise on the line that interferes with or contaminates every message.) In this incessant commerce, there is no Being of Language. But there are always voices: voices and more voices, voices within and behind voices, voices interfering with or replacing or capturing other voices.

I hear these voices whenever I speak, whenever I write, or whenever I pick up the telephone. Marshall McLuhan argues that technological change literally produces alterations in the ratio of our senses. The media are artificially generated parasites, prosthetic organs, "the extensions of man." Contemporary electronic telecommunications media are particularly radical, as they don't just amplify one sense organ or another, but represent an exteriorization of the entire human nervous system. Today we don't need shamans any longer, since modems and FAXes are enough to put us in contact with the world of vampires and demons, the world of the dead. Viruses rise to the surface, and appear not just in the depths of our bodies, but visibly scrawled across our computer and video screens. In William Gibson's *Count Zero*, the Haitian loas manifest themselves in cyberspace: spirits arising in the interstices of our collectively extended neurons, and demanding propitiation. In certain issues of the DC comic book *Doom Patrol*, written by Grant Morrison and illustrated by Richard Case, we learn that the telephone is "a medium through which ghosts might communicate"; words spoken over the phone are "a conjuration, a summoning." The dead are unable fully to depart from the electronic world. They leave their voices behind, resonating emptily after them. The buzzing or static that we hear on the telephone line is the sum of all the faint murmurings of the dead, blank voices of missed connections, echoing to infinity. These senseless utterances at once feed upon, and serve as the preconditions for, my own attempts to generate discourse. But such parasitic voices also easily become fodder for centralizing apparatuses of power, like the military's C³I system (command/control/communication/intelligence). *Doom Patrol* reveals that the Pentagon is really a pentagram, "a spirit trap, a lens to focus energy." The "astral husks" of the dead are trapped in its depths, fed to the voracious Telephone Avatar, and put to work on the Ant Farm, "a machinery whose only purpose is to be its own sweet self." As Burroughs also notes, the life-in-death of endless viral replication is at once the method and the aim of postmodern arrangements of power.

No moribund humanist ideologies will release us from this dilemma. Precisely by virtue of their moribund status, calls to subjective agency, or to collective imagination and mobilization, merely reinforce the feedback loops of normalizing power. For it is only by regulating and punishing ourselves, internalizing the social functions of policing and control, that we ever arrive at the strange notion that we are producing our own proper language, speaking for ourselves. Burroughs instead proposes a stranger, more radical strategy: "As you know inoculation is the weapon of choice against virus and inoculation can only be effected through exposure." For all good remedies are homeopathic. We need to perfect our own habits of parasitism, and ever more busily frequent the habitations of our dead, in the knowledge that every self-perpetuating and self-extending system ultimately encounters its own limits, its own parasites. Let us become dandies of garbage, and cultivate our own tapeworms, like Uncle Alexander in Michel Tournier's novel *Gemini* (*Les Météores*). Stylize, enhance, and accelerate the processes of viral replication: for thereby you will increase the probability of mutation. In Burroughs's vision: "The virus plagues empty whole continents. At the same time new species arise with the same rapidity since the temporal limits on growth have been removed. . . . The biologic bank is open." It's now time to spend freely, to mortgage ourselves beyond our means.

Don't try to express "yourself," then; learn rather to write from dictation, and to speak rapturously in tongues. An author is not a sublime creator, as Dr. Frankenstein wanted to be. He or she is more what is called a channeller, or what Jack Spicer describes as a radio picking up messages from Mars, and what Jacques Derrida refers to as a sphincter. Everything in Burroughs's fiction is resolved into and out of a spinning asshole, which is also finally a cosmic black hole. In Chester Brown's comic book *Ed the Happy Clown* (originally *Yummy Fur*), there is a man who suffers from a bizarre compulsion: he can't stop shitting. More comes out than he could ever possibly have put in. It turns out that his asshole is a gateway to another dimension, a transfer point between worlds. This other dimension isn't much different from ours: it has its own hierarchies of money and power, its own ecological dilemmas, even its own Ronald Reagan. But what's important is the process of transmission, and not the nature of the product. Waste is the only wealth, and that's how masterpieces are born. "Why linger over books to which the author has not been palpably *constrained*?" (Bataille). This constraint, this pressure in my intestines and bowels, marks the approach of the radically Other. It's in such terms, perhaps, that we can best respond to George Clinton's exhortation: "Free your mind, and your ass will follow."

2. The Insect People of Minraud

We all have our totem animals, our familiars, our spirit guides. They are usually other mammals, sometimes birds, occasionally even reptiles or amphibians; but they are almost always vertebrates of one sort or another. Our relationships with insects, on the other hand, tend to be stranger, more uncanny, more disturbing. Few of us—Spiderman aside—willingly accept intimacy with the arthropods. "Insect collecting is a hobby few can share," as Shonen Knife gently laments. Burroughs waxes lyrical about cats, about lemurs, about "sables, raccoons, minks, otters, skunks and sand foxes"; but he can only approach arthropods with an obsessive, fascinated repulsion. His novels are filled with hallucinatory visions of the insect- and centipede-ridden realms of Minraud and Esmeraldas, places of sexual torture and sacrifice. Exceptions to this horror can perhaps be made for the beauty of butterflies, and for the savoriness of certain non-insect arthropods, like crustaceans. But almost nobody enjoys our enforced proximity to bedbugs, cockroaches, and houseflies. Is our disgust simply the result of being confronted with a life form so utterly alien? Our lineage separated from theirs more than 600 million years ago, even before the Cambrian explosion. The insects' modes of feeding and fucking, those two most crucial biological functions, are irretrievably different from ours. Looking across the vast evolutionary gap, we are seized by vertiginous shudders of gastronomical nausea and sexual hysteria:

> We have all seen nature films in which enormously magnified insects unfeelingly dismember their prey. Their glittering multifaceted eyes stare at the camera while their complex mouthparts work busily, munching through still-struggling victims. We can empathize with our closer relatives the lions, who at least seem to enjoy their bloody work. But when the female mantis bites the head off its mate in order to release its copulatory reflex, it does so at the behest of an instinct that seems to have nothing to do with love, hate, or anything else to which we can remotely relate. (Christopher Wills, *The Wisdom of the Genes*)

Such an enthralled disgust is crucial to the postmodern experience of limits. The narrator of Clarice Lispector's *The Passion According to G. H.* is captivated by the sight of a wounded cockroach, trapped in a doorjamb as a "whitish and thick and slow" paste oozes out of its ruptured body.

After pages of obsessive contemplation and description, she ritually devours the cockroach, finding in it the impossible "embodiment of a prehistoric, pre-symbolic, ecstatic primal divine matter" (Camillo Penna). But this effort at communion necessarily fails. The flesh of the squashed bug is *sacred*, as Bataille might put it, because it is primordially ambivalent: it arouses both disgust and desire, at once demanding and repelling our intimate contact. We cannot touch, much less eat, this debased matter; and yet we cannot stop ourselves from touching and eating it. Insect life is an alien presence that we can neither assimilate nor expel. Professional exterminators know this well, and so do the best theologians and philosophers. Much ink has been spilled recently exploring Thomas Nagel's question, "what is it like to be a bat?"—or more accurately: is it possible for us to know what it's like to be a bat? But the whole discussion looks suspiciously like a replay of the old philosophical canard regarding the alleged unknowability of "other minds," only tricked out this time in postmodern drag. And in any case the bat is still a mammal, a fairly close relative of ours. That makes it all much too easy. Wouldn't it be more relevant and useful to pose the question of radical otherness in biological terms, instead of epistemological ones? It would then become a problem, not of metaphorically entering the mind of a bat, but of literally and physically entering—or metamorphosing into—the body of a housefly. And resolving such a problem would involve the transfer, not of minds, but of DNA. What's important is not to intuit what it might be like to be another species, but to discover experimentally how actually to become one. Such is the import of Cronenberg's film *The Fly*.

Burroughs cites Rule One of the basic biologic law, rigidly enforced by the Biologic Police: "Hybrids are permitted only between closely related species and then grudgingly, the hybrids produced being always sterile." To innovate means to violate this law, to introduce alien genetic material, to assume the risks of "biologic and social chaos." But then, viruses and bacteria are doing this all the time. There's nothing new about genetic engineering; as Lynn Margulis points out, humans are only now adopting techniques that prokaryotes have already been practicing for billions of years. As for viruses, they seem just to be transposable elements—such as can be found in any genome—which have revolted against the tyranny of the organism, or otherwise gotten out of hand. From meiosis to symbiotic merger, every genetic recombination is a new throw of the dice. No such process can be controlled or determined in advance. In Cronenberg's film, *Homo sapiens* meets *Musca domestica* only by the sheerest contingency. The transformation of Seth Brundle (Jeff Goldblum) into an insect—or more precisely, into the monstrous hybrid Brundlefly—is a statistical aberration: an improbable accident, a fortuitous encounter, an irreproducible, singular event. That's why Seth never quite comprehends what's happening to him, at least not at the moment that it happens. His scientific consciousness lags perpetually behind his ceaselessly mutating body. His theories about his condition are out of date by the time he utters them. Cronenberg's human-turned-fly is the postmodern realization of Nietzsche's prophecy of the Overman: "man is something that should be overcome." For the *Ubermensch* is not the "higher man," nor is he any sort of fixed entity. Rather he is a perpetual becoming, an ungrounded projection into unknowable futurity. The singular hybrid Brundlefly is just such a body, without stable identity, caught in the throes of transformation. Did Nietzsche ever suspect that his great metaphysical longing would be most compellingly realized in insect form? Any scientist can make observations about how flies (or bats, or humans) act in general; but even Seth Brundle never *knows* from the inside "what it's like" to be a fly. For "what it's like" necessarily involves the irreversible *othering* of the knower: the "going-under" of the Overman, the continual "becoming" of Brundlefly. The pursuit of knowledge, as Foucault puts it, should result not just in the "acquisition of things known," but above all in "the going-astray of the one who knows."

Insects are well ahead of humans in this regard. Radical becomings take place routinely in their own lives. This is especially so in groups that pass through pupal metamorphosis. Their bodies are broken down and completely rebuilt in the course of transmutation from the larval to the mature stage. Is the butterfly "at one" with the caterpillar? Is this housefly buzzing around my head "the same" as the maggot it used to be? One genome, one continuously replenished body, one discretely bounded organism; and yet a radical discontinuity both of lived experience and of physical form. The surplus value accumulation of larval feeding gives way to lavish expenditure: the extravagant coloration of the butterflies, the coprophilic copulation of houseflies and others. Insect life cycles continually affirm the possibilities of radical difference—even if ants and bees would co-opt this difference into the homogenizing mold of the State. Every insect is a "singularity without identity," in Giorgio Agamben's phrase. The fringe biologist Donald I. Williamson even goes so far as to argue that larval stages are remnants of symbiotic mergers between formerly independent organisms. But whether or not this be literally the case, Brundle's hybridization certainly opens the door to yet stranger metamorphoses. The body of an insect—far more radically than the mind of a dialectician—is perpetually "other than itself."

The high intelligence and adaptive flexibility of mammals is usually attributed to our premature birth, and our consequent long period of growth outside the womb. Genetics is supplemented by empirical learning and parental guidance. We lay down numerous memory traces, and build up complex personalities. Learning doesn't play such a role in insect development: not only because they have too few neurons to store all that information, but more crucially because memory traces cannot survive intact through the vast physiological changes of pupal transmutation. We higher mammals like to congratulate ourselves on our supposed ability to alter our own behavior adaptively in the span of an single lifetime. But this complacency may well be exaggerated. Innovation is harder than it seems. Insects usually manage to adapt to changed environmental circumstances a lot faster than we do, thanks to their greater propensity to generate mutations, and their far higher rate of genetic recombination over the course of much shorter reproductive cycles. In humans and other mammals, once memory traces are forged and reinforced, it's nearly impossible to get rid of them. And as if that weren't enough, we've also instituted *traditions* and *norms of critical reflection*, the better to police our identities, and to prevent our minds and bodies from going astray. Education, after all, is just a subtler and more sadistically refined mode of operant conditioning than the one provided by direct genetic programming. As Elias Canetti remarks, no totalitarian despot can ever hope to dominate and control his subjects so utterly as human parents actually do their children. We accept such discipline largely because we feel compensated for it by the prospect of imposing it in turn upon our own descendants. Our mammalian talents for memory and self-reflection serve largely to oppress us with the dead weight of the past. Morse Peckham is right to insist that only "cultural vandalism"—the aggressive undermining of established values through random, mindless acts of destruction—can free us from this weight, and stimulate social innovation. We humans need to push ourselves to such disruptive extremes; otherwise we have no hope of matching the insects' astonishing ability to adaptively alter their physiology and behavior in a relatively brief time. Unburdened by mammalian scruples, insects effortlessly practice the Nietzschean virtue of *active forgetting:* the adult fly doesn't remember anything the maggot once knew.

Postmodern biology is increasingly oriented toward what might be called an insect paradigm. In postmodern biotechnology, according to Donna Haraway, "no objects, spaces, or bodies are sacred in themselves; any component can be interfaced with any other if the proper standard, the proper code, can be constructed for processing signals in a common language." The organicism of romantic and modernist thought—together with its political correlate, the disciplinary "biopolitics" so powerfully described by Foucault—has given way to a new model of life processes. Postmodern bodies are neither "vitalistic" nor "mechanistic." They are structured through principles of modular interchangeability and serial repetition; they innovate, not on the basis of any pregiven criteria, but experimentally, by continual trials of natural selection. Arthropod body plans are especially postmodern, built as they are on multiply repeated segments, which can be fused or altered to generate new, differentiated structures. (The organic metaphors of the nineteenth century, in contrast, are idealizations of vertebrate body plans.) Genetic engineering, whether carried out in the laboratory or in "nature," requires just such a modular flexibility. Stephen Jay Gould, reflecting on the astonishing variety of arthropod forms discovered in the fossils of the Burgess shale, suggests that the initial Cambrian diversification of multicellular life progressed precisely in this way. Cambrian evolution seems to have taken the form of a "grab bag," mixing and matching body segments in a process much like "constructing a meal from a gigantic old-style Chinese menu: one from column A, two from B, with many columns and long lists in every column" (*Wonderful Life*). This kind of thing doesn't much happen in macroevolution any longer; but it's still crucial on the molecular-genetic level, as Christopher Wills argues in *The Wisdom of the Genes.* Certain mimetic butterflies, for instance, have linked "supergene complexes" that allow them alternatively to mimic any one of a number of vastly different model species. Segmented repetition with modular variation remains the basic organizing principle of all insect genomes: hence the frequency of homeotic mutations—multiplied wings and legs, antennae transformed into legs, added or subtracted segments—in laboratory strains of *Drosophila.* Melancholy old conservatives like Jean Baudrillard fear that postmodern modular coding leads to a preprogrammed "satellitization of the real," and finally to its total "extermination." But even the slightest acquaintance with insects will convince you that—contrary to Baudrillard's claims—"the hyperrealism of simulation" allows for a far greater explosion of change, multiplicity, and sheer exuberant waste than traditional organicist models of production and circulation ever did.

Haraway points out that recent developments in postmodern biology involve a radical problematization and "denaturalization" of all notions of the organism and the individual. Witness Lynn Margulis on the symbiotic basis of eukaryotic cells, Richard Dawkins on "selfish genes," parasitism, and the "extended phenotype," and Leo Buss on the multiple, variant cell lineages of mammalian immune systems. When we look at the molecular-genetic basis of life, all we can find are differences and singularities: multiple variations, competing alleles, aberrant particle distributions, unforeseeable sequence transpositions. These multiplicities never add up to anything like a distinct species identity. Postmodern biology thus deals not with fixed entities and types, but with recurring patterns and statistical changes in large populations—whether these be populations of genes or populations of organisms. It tends to emphasize anomalous phenomena like retroviral infections and horizontal gene transfers; in such encounters, alteration "ceases to be a hereditary filiative evolution, becoming communicative or contagious" (Deleuze and Guattari). Postmodern biology moves directly between singularities without identity and population multiplicities, without having recourse either to intervening, mediating terms, or to overarching structural orders. It rejects the "holism" formerly attributed both to the individual organism and to the larger ecosystem. Look at the mutations and transpositions haunting any genome, or observe the behavioral quirks of the cockroaches invading your apartment. You will find what Deleuze and Guattari call

> molecular, intensive multiplicities, composed of particles that do not divide without changing their nature, and distances that do not vary without entering another multiplicity and that constantly construct and dismantle themselves in the course of their communications, as they cross over into each other at, beyond, or before a certain threshold. (*A Thousand Plateaus*)

The obsolescence of those old organicist and holistic myths opens the way to strange new social and political arrangements. In our postmodern world, the "disciplinary power" analyzed by Foucault is continually being displaced into more subtly insidious modes of oppression. The ubiquitous codes of an "informatics of domination" (Haraway) are initially deployed by government bureaucracies, and then "privatized" as the property of multinational corporations. Such flexible and universal codes, insinuating themselves within all situations by a process of continual modulation, are the hallmark of what Deleuze, following Burroughs, calls the postmodern "society of control." Cybernetic regulation is the human equivalent of the pheromone systems that regulate all activity in an ant colony. But let's not assume that this new arrangement of power forecloses all possibilities of resistance and change. As Deleuze says, "there's no need to fear or hope, but only to look for new weapons." Seth Brundle speaks of his paradoxical desire to become "the first insect politician," suggesting the possibility of an alternative insect politics, different from the totalitarianism of ants and bees. Consider that flies, like midges and mosquitoes, tend to swarm; and that locusts periodically change form, and launch forth into mass nomadic rampages. Such insects form immense crowds without adopting rigidly hierarchical structures. Their loose aggregations offer far more attractive prospects for postmodern sociality than do the State organizations of the Hymenoptera. Insect swarms are populations in continual flux, distributing themselves randomly across a vast territory. They are altered by the very processes that bring them together, so that they can neither be isolated into separate units, nor be conjoined into a higher unity. "Their relations are distances; their movements are Brownian; their quantities are intensities, differences in intensity" (Deleuze and Guattari). If postmodern power is exemplified by the informational feedback mechanisms of the "insect societies," then maybe a postmodern practice of freedom can be discovered in the uncanny experience of the insect swarms. The next time you see flies swirling over a piece of dung, reflect upon what Agamben calls the "coming community," one that is not grounded in identity, and "not mediated by any condition of belonging"; or upon what Blanchot calls "the unavowable community" or "the negative community" or (quoting Bataille) "the community of those who do not have a community."

Postmodern politics, like postmodern biology, must in any case come to grips with natural selection. The romantics and the modernists alike misconceived evolution in melioristic or moralizing terms. Even today, New Age sentimentalists search frantically for any metaphysical solace that might palliate the harshness of neo-Darwinian struggle. We hear tales of beneficent feedback mechanisms (Gregory Bateson, James Lovelock), of heartwarming cooperative endeavors (Francesco Varela, Stephen Jay Gould), of synchronic species progression (Rupert Sheldrake), or of strange attractors at the end of history (Terence McKenna). These are all visions of a world without insects, one in which change would always conform to our petty bourgeois standards of niceness and comfort. Burroughs and Cronenberg know better, as do biologists like Richard Dawkins. We live, as Burroughs reminds us, in a "war universe." If we want to survive, we must avoid the facile self-deceptions of teleological and rule-driven explanations. Let us rather construct our "war machines" according to pragmatic, immanent, selectionist principles. Mammalian immune systems in fact already work in this way: they "learn" to recognize and destroy enemy proteins as a result of differential reproduction rates among widely varying T cells. Similar models for the adaptive growth of neurons in the human brain—"neural selectionism" and "neural Darwinism"—have been proposed by Gerald Edelman, Steven Pinker, and others. And artificial intelligence research has started to explore the possibilities of allowing selectional processes to operate blindly, instead of imposing predetermined algorithms. All such selectional systems are what Deleuze and Guattari call *desiring machines* or *bodies without organs:* they are not closed structures, but relational networks that "work only when they break down, and by continually breaking down." Breakdowns are inevitable, since the process of adaptation is never rapid enough to keep up with the pace of continual change. And every breakdown brings to the fore an immense reservoir of new, untapped differences and mutations: material in random variation upon which selection can operate. These selectional processes, therefore, do not guarantee us anything in advance. They do not provide for a future that will comfortingly resemble the present or the past. They do not help us to imagine how things might be better—that old utopian fantasy, much beloved of "progressive" social critics. Rather, their political efficacy lies in this: that they actually *work* to produce differences we could not ever have imagined. They provoke innovations far stranger and more radical than anything we can conceive on our own. "I love the uncertainty of the future," as Nietzsche so stirringly wrote.

So cultivate your inner housefly or cockroach, instead of your inner child. Let selectional processes do their work of hatching alien eggs within your body. And don't imagine for a second that these remarks are merely anthropomorphizing metaphors. We can kill individual insects, as spiders do; but we can't for all that extricate ourselves from the *insect continuum* that marks life on this planet. The selectional forces that modulate insect bodies and behaviors are also restlessly at work in our own brains, shaping our neurons and even our thoughts. Does such an idea revolt you? The problem might be that we can't read insect expressions: we don't know what they are thinking, or even if they are thinking. But this may just be an unwarranted vertebrate physiological prejudice; after all, "insects are naturally expressionless, since they wear their skeletons on the outside" (Christopher Wills). Watch for when the insect molts, and its inner vulnerability is exposed.

We should reject all distinctions of inner and outer, as of nature and culture. How could you ever hope to separate genetic influences from environmental ones, or biology from sociology? Those social critics who think "biological" means ahistorical and unchanging—and reject naturalistic explanations on that basis—clearly don't know what they are talking about. The bizarre, irreversible contingencies of natural history and cultural history alike stand out against all endeavors to endow life with meaning, goal, or permanence. Entomology is far less essentialistic, far more open to difference and change, far more attentive to the body, than are, say, cultural critiques grounded in Frankfurt School post-Marxism and Lacanian psychoanalysis. It's common in well-meaning academic humanist circles to loathe and despise sociobiology. But this isn't just a matter of disputing some rather dubious claims about particular aspects of human behavior. What many of my colleagues really can't forgive is sociobiology's insistence upon biological embodiment itself. It's not really a question of whether this or that gender trait is really "written in our genes," so much as it is a case of the panicky denial of evolutionary contingency, or genetic limitation, altogether. "Dialectical" biologists like

Richard Lewontin, together with their social-determinist allies, merely perpetuate a massive, and quite traditional, idealization of human culture: one that has long fueled delusive fantasies of redemption and transcendence, and that has served as an alibi for all sorts of controls over people's lives, and moralistic manipulations of actual human behavior. Edward O. Wilson, to the contrary, made only one real mistake when he came to systematize the discipline of sociobiology: this was his choice of ants, rather than houseflies or cockroaches, as an implicit reference point for examining "human nature." Be this as it may, entomological intuitions continue to be more illuminating and provocative than narrowly humanistic ones. Maurice Maeterlinck well expressed the uncanny fascination of insect life nearly a century ago: "The insect brings with him something that does not seem to belong to the customs, the morale, the psychology of our globe. One would say that it comes from another planet, more monstrous, more dynamic, more insensate, more atrocious, more infernal than ours." What has changed in this picture in the last one hundred years? Only one thing. We postmoderns have come to realize that such alien splendor is precisely what defines the cruelty and beauty of *our* world.

PART II:

SOME GENDERS

THREE

The End of the World of White Men

Kathy Acker

It was the days when men were cutting their cocks off and women were putting on strap-ons . . .

The Muses sometimes form in these low haunts their most lasting attachments . . .

Artaud speaks:

When O was a young girl, above all she wanted a man to take care of her.

In her dream, the city was the repository of all dreams.

A city which is always decaying. In the center of this city, her father had hanged himself.

This can't be true, O thought, because I've never had a father.

In this dream, she searched for a father.

She knew that it was a dumb thing for her to do because he was dead.

Since she wasn't dumb, she thought, she must be trying to find him so that she could escape from the house in which she was living, which was run by a woman.

O went to a private detective. He called O a dame.

"I'm looking for my father."

The private eye, who in one reality was a friend of O's, replied that the case was an easy one.

O liked that she was easy.

And so they began. First, according to his instructions, O told him all that she knew about the mystery. It took her several days to recount all the details.

At that time it was summertime in Dallas. All yellow.

O didn't remember anything in or about the first period. Of her childhood.

After not remembering, she remembered the jewels. When her mother had died, a jewel case had been opened. The case, consisting of one tray, held insides of red velvet. O knew that this was also her mother's cunt.

O was given a jewel which was green.

O didn't know where that jewel was now. What had happened to it. Here was the mystery of which she had spoken.

The private eye pursued the matter. A couple of days later, he came up with her father's name.

"Oli."

The name meant nothing to her.

"Your father's name is Oli. Furthermore, your father killed your mother."

That's possible, O thought, as if thinking was dismissing.

The detective continued to give her details about her father: he was from Iowa and of Danish blood.

All of this could be true because what could she in all possibility know?

When O woke up out of her insane dream, she remembered that her mother had died eight days before Christmas. Despite the note lying beside the dead body in which the location of the family white poodle was revealed, the cops were convinced that the mother had been murdered. By a man unknown. Since it was now Christmas, these cops had no intention of investigating a murder rather than returning to their families, Christmas warmth, and holiday.

O realized, for the first time in her life, that her father could have murdered her mother. According to the only member of her father's family whom she had ever met, a roly-poly first cousin whose daughter picked up Bowery bums for sexual purposes (according to him), her father had murdered someone who had been trespassing on his yacht.

Then, the father had disappeared.

O became scared. If her father had killed her mother, he could slaughter her. Perhaps that's what her life had been about.

During this period of time, O lived and stayed alive by dreaming. One of the reveries concerned the most evil man in the world.

It was at a fancy resort that was located in the country, far from the city: O stood on one of the disks, as if on a giant record, which jutted out of a huge cliff. Shrubbery was coming out of parts of the rock. Each record lay directly over and under another record, except for *the top* and *the bottom.* The one on which O was perched thrust further into a sky which was empty; the record was a stage.

In the first act of this play, O learned that evil had entered the land. That the Father, who is equivalent to evil, is appropriating or thieving all of his son's possessions with success. Both the father and the son were standing behind O. The father began to increase and deepen his evilness by torturing his son. He inflicted pain solely physically. O actually saw the older man point at her three different machine guns, each of which was different from the other two. O understood what this man was saying to her: he wanted to scare, rather than to shoot her.

Then he laughed.

O hated the man of evil as much as it was possible for her to abhor anyone.

Either the next day or an unknown number of days later, while O in her classroom was doing whatever she did, which, according to her contract with the university, had to be called *teaching*, but teaching was something which she didn't understand, she noticed that her students were no longer paying attention to whatever she was doing and were whispering to each other. Worse, they were staring at what she didn't know.

In the center of her classroom, there was a cat chasing a rat. The rat, for O, was in the center, was almost touching her. Then the rodent leaped straight up: it was closer to her. O couldn't understand why she wasn't terrified. Why wasn't she jumping up on the seat of a wood chair and lifting the front of her skirt up over her cunt? Artaud wrote: we must get rid of Mind as we must get rid of all literature. But instead of doing *that which she was supposed to do*, perhaps because she identified with cats, she kicked this white rat in the shins. Of course, it was by accident. As she was kicking hard, she saw that the animal was a smaller version of the huge stuffed white cat with whom she slept every night.

O needed to be held by her cat in order to fall asleep.

Artaud continued: I say that Mind and life interconnect at all levels. I would like to make a Book to disturb people, like an open door leading them to where they have never gone. Simply a door communicating with reality.

O saw that the cat or rat wasn't going to die.

Then her real search began. She had to find the torturer so that she could get rid of all the evil that was in the world. His son and O had become partners and mercenaries: it was he who taught O that she would be able to search only if she got rid of her fear of evil.

For some reason unknown to O, she was frightened of everyone and everything.

The father of evil had left them a clue to his whereabouts: D.N.

Nobody seemed to know whether D.N. were the initials of someone or something or whether the letters were part of a language nonapprehendable by reason. O and the son thought that *D.N.* was the name of a coffee joint . . .

. . . .

. . . . .They came to a deserted western street. The coffee joint found in this loneliness named "a street," within all this yellow, wasn't named *D.N.* or anything . . .

. . . .

. . . . .They came to a ranch. The main building, at first not noticed or noticeable, was a white one-story, peeling paint. To its right, in one of its sides, a café-in-the-wall.

A girl fed her dog-horse, large as a large horse, a plate of raw hamburger. She had once been married to the man of evil's son. Now, separated from him and living on the ranch, she was happy.

This clue which was true informed O and the partner that the man of evil was present.

He walked up to her. In all that openness. There was no one but those two. Then O realized that all that had happened to her, the narrative, had happened because and only because she was attracted to this man. The evil father. She hated him because he was irredeemably violent.

O began to teach him that he could transform his violence into mutual pleasure through sex.

It was at this point that O decided that she wanted to go where she had never been before . . .

O speaks:

The revolution had yet to begin in China. The word *revolution* no longer meant anything to us because the same governments now owned everything. There was nowhere left to go. Wherever we were living, all of my friends, including me, were dying before we reached old age and, before that, living in ways that crossed social and other limits because otherwise living was unbearable.

I no longer had interest in politics.

I had come to China as I usually came: I was following a guy.

I believed that we were in love.

It didn't matter the name of this unknown city. All unknown cities, in China, hold slums which look exactly like each other: each one a labyrinth or an actual dream in which streets wind into streets which are winding into more streets and every street goes nowhere. Perhaps, because all the signs have disappeared.

The poor eat whatever they can get.

Right before the revolution, the Chinese government told its people that the recession was over; the poor could no longer distinguish between economic viability and disability. Some of them walked around with needles sticking out of their bodies.

Most of the females had whored for money.

In China, my boyfriend, W, told me that if I loved him, I would whore for him. I think that also W gets off on women who are whores. I don't know whether or not he had deep feelings for me and what those feelings were. I wonder, again and again, why I run after men who don't have any feelings for me and why I, for this reason, don't have a boyfriend.

It is my mother, rather than my father whom I have never known, who dominates my rational life. When she was alive, my mother didn't notice and, when she noticed, hated me. She wanted me to be nothing or something worse because my appearance in her womb, not yet in the world, had caused her husband to leave. What my mother who was ravishingly beautiful, charming, and a liar told me. While she was alive.

Absence isn't the name only of my father. Every whorehouse is childhood; the one into which W placed me was a hounfor named *Ange.*

In the outside world, which is the one outside a whorehouse, men fear women who are beautiful and run away from them: a ravishing woman who's with a man has another kind of deep wound somewhere else. My mother was weak in this way and her weakness is now my fate.

Whereas inside the brothel, the females, whoever they physically are, must always appear beautiful to men; they have been imprisoned in order to fulfill men's fantasies. In this way what was known as "the male regime" separated its fantastical life from its rational existence.

Since I was the first white girl who had ever entered this hounfor, the others, including the madam who had once been a male, hated me. They sneered at my characteristics, such as my politeness; what they really detested was that necessity, economic necessity, wasn't driving me into prostitution. To them, the word "love" meant nothing. But I wasn't becoming a whore because I would do anthing for W, anything to convince him to love me, a love I was beginning to know I would never receive. I had entered the brothel willingly so that I could become nothing because, only then, could I begin to see.

I had had no idea what I was doing.

I entered the brothel and the madam took away all of my possessions, even my tiny black reading glasses. It was as if she was a prison matron. She said that because I was white, I thought that I deserved things such as to be happy, that I deserved to possess commodities. Such as happiness. That I was too pale, delicate to be able to bear living in this place.

The girls thought that I could leave the cathouse whenever I wanted. But I couldn't walk away because inside the whorehouse I wasn't anybody.

There was nobody to walk away. I was now a child: if I rid myself of childhood, there would be nothing left of me.

Artaud understood this.

Later on, the girls accepted me as a whore. As they were. Then I started to wish that I would love a man who loved me.

In their spare hours, the whores visited fortunetellers. There were many prescients in the slum. Though I soon started accompanying my co-workers, I was scared to say anything to these women who had once been in the business. I would stand in those shadows and rarely ask anything: for I didn't want to tell anything about myself. When I did inquire about a future, I asked as if there were no such thing. I only felt safe asking about the details of daily life, johns, and defecation, all that was a dream.

As if dreams aren't real.

Fortunetellers wandered around the streets right outside the hounfor.

The fortune, mine, which I remember was based on the card of the Hanged Man:

The woman who was reading the cards still took tricks.

"Does that mean that I'm going to suicide?"

"Oh, no, O. This card says that you're a dead person who's still alive. You're a zombie."

I knew better. The Hanged Man or Gerard de Nerval is my father and every man I fuck is him. Like I said, fortune is dead whenever all the men are hung.

My father is the owner of the cathouse. He's sitting in his realm of absence and he surveys all that is not.

The cards clearly show me that I hate him. When a message travels from the invisible to the visible, the messenger is an emotion. My anger, this messenger, will lead to revolution. Revolutions are always dangerous.

The cards said worse. They told us, whores, that the revolution which is just about to happen, due to its own nature or origin, must fail. When it fails, when sovereignty be it reigning or revolutionary has finally disappeared, when sovereignty eats its own head as if it's a snake, when the streets are again dust and decay but a different dust and decay, all my dreams, which are me, will be shattered.

"It's then," the slut-fortuneteller said, "that you'll find yourself on a pirate ship."

What cards I remember told me my future is freedom.

"What'll I do when there's no one in the world who loves me? When all my existence is this not-ocean or freedom?"

The cards proceeded to give images of stress, illness, disease . . .

Whores are diseased. This is why no one loves them. I had now been in the cathouse for a month. W hadn't once visited me; he had never cared about me. I was a whore because I was alone.

Three helpers were going to show me how to become free. They were a cock who was the largest of all cocks, the journey into the land of the dead, and Yemaya.

I was trying to get rid of loneliness and nothing will ever rid me of loneliness until I suicide.

Artaud speaks:

O said, I want to go where I have never been before.

I was living in a room that was in the slum. I was still sane.

I was just a boy. All I saw was the poverty of those slums. In order to counteract the poverty that was without, and within me, I ran to poetry. Especially to the poetry of Gerard de Nerval who wanted to stop his own suffering, to transform himself, but instead hanged himself from a rusty picture nail.

I had no life. I only loved these poets who were criminals. I began to write letters to people whom I didn't know, to those poets, not in order to communicate with them. To do something else. I wanted to hang myself.

Dear Georges, I wrote.

I have just read, in *Fontane* magazine, two articles by you on Gerard de Nerval, which made a strange impression on me.

I am a limitless series of natural disasters and all of these disasters have been unnaturally repressed. For this reason, I am kin to Gerard de Nerval who hanged himself in a street alley during the hours of a night.

Suicide is only a protest against control.

Artaud.

The alleyways were lying all around me. They ran every which way so haphazardly that they stopped. Here lay the hounfor.

I would watch man after man walk through the doors. Men went to the brothel, not in order to have the sexual intercourse they could have on the outside, but to enact elaborate and torturous fantasies which, one day, I'll be able to describe to you.

I'll be able when there is pleasure in this world. At that time I did not have a lover nor did I know what it was to be a body.

Day after day I would look through one of my windows into one of theirs. It was there that I first saw O who was naked. My eye would follow her, as much as it could, so that it could clear away all that was before and behind her.

I would die for her. Whenever a man hangs himself, his cock becomes so immense that for the first time he knows that he has a cock.

One day, O emerged from the brothel. I saw her stand on the edge of the doorway and look away from the brothel. Obviously she was terrified. Finally, one of her feet peeped over the doorframe's bottom. I had no idea what was mirrored in those eyes. Her feet moved three times back and forth across the doorstep.

As soon as she was fully outside, she began to turn in the same ways that winds move through the airs. Perhaps she was meeting this outside and the air for the first time. Perhaps, in the stale air of the brothel, O was a "she" and now O was another "she" who was indistinct from "air." I watched her begin to breathe. It was in this way that O, alien, encountered poverty, the streets which my body were daily touching. The streets whose inhabitants ate whatever they could and then, when they could no longer eat, committed suicide.

The streets reminded O of her childhood. When she had been a child, she had always been alone. Even though she'd had a half-sister who was now married to a European armaments millionaire. Every summer, O's mother, so that she would never have to see her oldest daughter, sent her to a posh summer camp. The camp was composed only of girls.

While the girls passed through the latest dances in each other's arms in the hour before they were ordered into dinner, O stood on the sidelines and watched. All she knew was that she couldn't dance because she wasn't like all the others. In the whorehouse, perhaps for the first time in her life, O had become safe because there were no humans. Not the men who visited her.

In the hounfor, she was naked.

Now that O knew safety, she had the power to return to her childhood. To the poverty that was mine. I watched O walk down street after street, searching for a body. I knew that when O had a body, she would belong to me.

O speaks:

Just after the first time W and I slept together, I knew that he didn't love me. I didn't know why. This area of not knowing or nausea left me shreds of belief to which I could cling and I clung, belief that in the future W might start to love me. Like a child who cannot believe that her mother doesn't care about her.

I remained in that brothel. One day W returned and told me that now he wanted me to meet the woman whom he adored even more than his own life. In order to do this he was going to take me out of the brothel for the day.

They had been together many years, before he ever met me. He informed me. That then she left him. It had been his fault: he wasn't a good human being. That she returned to him, in China, and now he wanted to be as good to her as it was possible for any human to be.

Though she had come back to him, she was still unsure whether she wanted to be with him, and that made him love her more.

I didn't know where I existed in W, therefore why he was telling me about the one woman whom he worshipped.

Maybe I could cling to this nausea. Maybe nausea is something. His fat body. I followed him into the streets outside the brothel. Into those streets which I had started to explore by myself. A bird flew through the sky.

His girlfriend was white like me. She was beautiful and rich. As soon as I met her, I knew that I didn't exist for her in the same way that I didn't exist for W, that she didn't know how to love. She was one of the owners. She was beautiful and rich. She had an identity.

I could love W which she never could, but did he want that? Did he realize all that I would be able to give him?

After the high-class dinner, he took his girlfriend and me back to the brothel and he tied me to my bed. Needles inserted into the flesh just below the lower lashes kept the eyes open. In front of me, W made love to this rich bitch first with his fingers. Delicately playing with her thick outer labia. Slowly they turned from pale pink to almost blood red. And opened to my eyes as the fingers went up. Some of his fingers were in her mouth. He was bending her over and then he turned around, her cunt juice dripping so much that I could see it from his tips, and put his long cock which I can still know in my mind into that cunt that must have been open, wanting, screaming for pleasure, whether she loved him or not, she was being fucked inserted thrust into pummelled bruised and all that comes out is pleasure, the body is pleasure, I have known pleasure, and I am watching the endless pleasure, as it comes again again again, that I have known and now I am being refused.

She, rich, can never know what my nausea, my lack is: what my pleasure is and so I am changing.

Throughout all of the dinner and the sex I was forced by myself to watch, I was wearing the deep red lipstick which color my mother wore. My mother had always walked around the house naked, touching her own white body; there she wore her menstrual blood on her mouth. There are no men: my father had left her before I was born.

Since I have never known you, every man I fuck is you. Father. Every cock goes into my cunt which is now a river named Cocytus. I have said that I will now only tell the truth: When you, Cock of all Cocks, you the only lay in the world, and I am the one who lives if not dies for sex, when you took a leave of absence skipped out ejaculated disappeared and vanished before I was born, you threw me, and I hadn't yet been born, into even another world.

The name of the world was China.

Who can understand China's teeming populaces, its children, its marching, student soldiers?

Artaud speaks (the rewriting of his first letter):

I am a violent being, full of fiery storms and other catastrophic phenomena. As yet I can't do more than begin this and begin again and again because I have to eat myself, as if my own body is food, in order to write. I don't want to write (talk the only way I know how) about myself. I want to discuss Gerard de Nerval. He made living: a living world: he made a living world out of myth and magic. The realm of myth and magic that he contacted was that of a Funeral. His own death and funeral.

I'll talk about death, my death, later.

The Tarot card in the realm of Nerval is The Hanged Man. Heidegger turned away from Hitler, reversed himself, and at the same time explained that "the very possibility of taking action" or "the will to rule and dominate" was "a kind of original sin, of which he found himself guilty when he tried to come to terms with his . . . past in the Nazi movement." Heidegger began at that time to emphasize, instead of Dasein, Sein, or an essentially reverent contemplativeness that might keep open the possibility of a new paganism in which no sovereignty could arise, no sovereignty out of the ashes of Hitler's aborted revolution.

Such contemplativeness is the hanged man in the realm of Nerval. Such contemplativeness is a human who's in the act of doing nothing or only turning inside-out, reversing, travelling the road into the realm of the dead from which he's returning alive. In other words, The Hanged Man card represents the slight possibility that this intolerable society (in which identity depends on possessing rather than on being possessed), this society in which I'm living, could change.

Gerard de Nerval was a sailor who descended into oblivion and, as he did, wrote against oblivion. He hated his own cock and so descended into the Cocytus, into oblivion, three times until his cock floated bloody on its waters. Or he hanged himself.

(O speaks: I spent day after day walking the streets. Looking for W whom I knew I would never again find.)

I am Gerard de Nerval who hanged himself 12:00 p.m. on a Thursday by his own hands; the other one died in Paris or he announced that this death would happen, he announced that he had died from loneliness coming from social rejection.

I, Gerard de Nerval, who wrote in the teeth of the utilitarian concept of the universe, hanged myself from an apron string tied to a grating. There was nothing left.

Now I, Gerard de Nerval, want to talk about the distance between hanging and The Hanged Man. I lived in an unbearable society and I murdered myself. Then a sailor, I journeyed to the lands between daily life and meaning (symbol). I AM MAD AND SO NO LONGER WILL MAKE SENSE. I hanged myself in order to scream more loudly; I know that is why suicide happens; and I died and my scream grew and castrated me so that this suffering turned into everlasting suffering. The name of everlasting suffering is *contemplation.*

In this way I turned myself inside-out:

Only the head can cut the head off. For instance, de Sade's valet, Latour, flagellated de Sade who hated de Sade. De Sade ran away from his own class. But I Antonin Artaud Gerard de Nerval am not running away: I am welcoming all the pain I can get because pain is the body.

I, Antonin Artaud, have hanged myself and I haven't died.

I'm living in a slum in China and I'm entering into sexuality. I'm now a hole so all the liquids can gush through me. I am sexuality. Being a man or an abyss of a man, I'm protecting my mother in the full knowledge that she hates my guts. My mother's a betrayer because she is beginning to know that she's God: she's perceiving her cunt and ovaries and all of her other holes to be the labyrinth and sacred. My mother is my holiness.

And I have adopted the embraces of Satan.

Antonin

O speaks:

Without W, I no longer want to be a whore.

Artaud speaks:

I entered the voodoo house so that I could meet O. It wasn't just a house. Its trees, small illness rooms, altars, and accompanying chambers called "ghuevos" were dedicated to the appearance of spirits in human bodies, to the spirits' possessions of human bodies.

Though the insides were complicated, staircases appearing out of nowhere and leading to floors not before seen, rooms upon rooms all in a jumble, somehow I knew exactly where O's bedroom would be. White doves, pigeons, chickens, precisely and delicately marked guinea-hens walked and flew out of these windows. Almost all of the inside walls were brilliantly whitewashed except where a certain spirit was being honored. On the ground floor, the madam stopped me to ask where I was going: I said that I was going to serve O.

She told me that I had to give money so I could be with O. Because I don't have any money, I was thrown out of the whorehouse.

I found myself in a marketplace where everything is being sold for everything else. There some of the poor didn't have any limbs. Others would do anything sexually for anyone. The children often said that, every harvest, a third of them were going to die if the growth wasn't bountiful. I decided that I had to stop the hell in which I was living.

I knew that they had thrown me out of the whorehouse because I didn't want to give O money, I wanted her to love me.

The denial of my sexuality planted in me the seeds of rebellion. There must be other women and men like me in that slum. Ones who would do whatever had to be done in order to change everything.

I looked for them in the holes, in the shadows that had eaten themselves up.

It was at this time that the revolutionaries, both male and female, met in what light came from the quarter moon.

The revolutionaries, who were mainly young, talked. Their city was a dump and growing. The only answer is that we get our hands on weapons.

"We're poor."

"A white man just gave us some money, probably in order to save his own neck."

Though I had no interest in weapons, I agreed to undertake the machine-gun delivery, dangerous at least, in return for the exact amount of cash I needed to buy O so that I could give her her freedom from the brothel.

I cut off my cock and blood out of a heart I had never known started to flow.

O speaks:

How long will this reign of masochism continue?

Artaud speaks (the letter continued, addressed directly to O):

Everywhere he went, de Nerval would take with him a scummy apron-string which had once belonged to the Queen of Sheba. De Nerval told me. Or it was one of the corset-strings of Madame de Maintenon. Or of Marguerite de Valois.

From this apron-string which was tied to a grating, he hanged himself.

The grating, black fetid stained with hound excretion and partly broken, was located at the bottom of the stone stairs which lead to the rue de la Tuerie. There's a straight drop from that stair platform downward.

As de Nerval swang there, a raven hovered over, as if it were sitting on his head, and cawed repeatedly, "I'm thirsty."

I'm thirsty are probably the only words that the old bird knows.

I, Antonin Artaud, though slum, am an owner: I own the objects and the language of suicide.

Why did Gerard de Nerval hang himself from an apron string? Why is this society which is China insane?

To learn why Gerard madly offed himself, I will enter his soul.

Gerard was a man like me. Women think that both of us are good-looking. Nerval wrote this:

. . . le dernier, vaincu par (Jehovah)
Qui, du fond des enfers, criait: "O tyrannie!"

Gerard was le dernier because he was just about to suicide; he was talking to God the tyrant whose very existence was putting Gerard in hell. So, Gerard suicided because of the existence of God; Gerard opposed the tyrant God by cutting off his own head. God is the head, le genie. He cut off his own head with a woman's apron-string so that he now has a hole between his arms so now he is a woman and can no longer tolerate the phallus-head order. This hole is the hole of nothingness. The soul of Gerard de Nerval has taught me that nothingness is the abyss of horror out of which consciousness always awakes in order to go out into something to exist.

The hole of the body, which every man but not woman including Gerard de Nerval and myself has to make, is the abyss of the mouth.

I have found my language which is why I can write this letter to you. O. You're naked. Gerard gave me a language that doesn't lie, that language that is spurting out of the hole of my body.

You're naked so I know you've got a body.

When Gerard cut off his cock, he made all that was interior in him exterior: all that is interior's becoming exterior and this is what I call a revolution and those who are holes are the leaders of the revolution.

I have gotten to know Gerard de Nerval and he was a revolutionary both before and after he hanged himself from an apron-string. He hanged himself from a woman's string in order to protest against social control because all suicide is a protest against control. I repeat that. After he castrated himself, language like screams came pouring through him.

I am Gerard de Nerval after he castrated himself because consciousness in the form of language is now pouring out through me and hurting me and I'm entering into sexuality. I want to own you, O.

There's only gloom or nothingness around what's rising out of me out of the nothingness; the gloom is everywhere, the streets made up of poverty, shadows of revolutionaries. I, Antonin Artaud, have put down the language that spouted out, I've written to O, but this language isn't me.

O speaks:

I have a fantasy, a sexual desire, which isn't a dream. First of all, there has to be a man. Then, there has to be sex between me and the man. These are my prerequisites for desire.

First, the man rejects my sex. With rejection, absence, desire awakes. Lust need memory or desire torment my body so badly that I become sick and am on the edge of death.

Just *when almost all is lost* (the world), this man returns to me. Taking me in his arms, he restores my life.

For the first time I knew that W would never love me. I was still living in the whorehouse. The dinner with him and his rich girlfriend was over. W would never come back to me, wrap his arms around me, and take me out of the brothel.

Knowledge that he would never love me was recognition that he never had.

Since I was no longer safe in the brothel, in this realm of fantasy, I became very sick. I hovered at death.

It was at this time that the student revolutionaries, armed more professionally than any of the cops around them, burst into the English embassy which was located in the section next to the slums. Paying in violent injury and death, they successfully annihilated the government building. When my health returned, I learned that W partly owned the cathouse. I had known that he was rich. I didn't care what he had felt or would feel about me: all I wanted was for him to be absent from me.

I wanted W to remain absent from me: I wanted nothing to change.

I learned that it had been W who had first given the terrorists the money to buy the weapons. Perhaps he hadn't know why. Perhaps there was a need in him to disrupt and to destroy. I didn't know W and I don't. When the revolutionary raid on the English had succeeded, red, probably he had become frightened. *For the first time in his life*, he had realized that to be rich and white is to be vulnerable. He understood that he was vulnerable. So when the revolutionaries had returned to him and asked for more funds, he had refused.

They began to beat him up. They almost killed him.

As soon as I learned that this had happened, I stopped hating W for not returning my love.

In the skirmish prior to the explosion of the English embassy, the young boy who had run guns to the revolutionaries had one of his arms severely injured.

With the other hand holding the money that he had earned by working for the terrorists, he entered the brothel. He found the madam and gave her the amount she had requested as the price of my purchase.

I knew nothing about the purchase of my freedom.

the degeneration of all my work

Behind my bedroom door, Artaud said that he had come back to me.

I replied, "I'm still sick: I don't want to see anyone."

I'm writing the way one dreams.

Then he forced himself into my room so I hit him. He fell down to the floor on the arm that had been broken. When he cried out, I was surprised.

"You're just a boy so how could you be hurting so badly?"

He told me that he had broken his arm in order to get the money to be able to buy me.

His arm was bent the wrong way for a human.

I understood that someone could hurt more than me. I reached down and lifted up his body, onto my thigh, as much as I was able. I only wanted to fuck him. At that point, pain was the same for him as sexual pleasure. For me, every area of my skin was an orifice and each part of his body transforming into an instrument could do and did everything to me.

We wondered at our bodies.

Artaud wrote in another letter:

I entered into sexuality and three times I became a hole through which liquids rushed, then poured. Three times I was plunged into the waters of oblivion.

Afterwards, when I saw O, I wanted to protect her because she worships her cunt.

O speaks:

I never saw Artaud again.

Weakened not only by the beating but then by the desertion of his rich girlfriend, W must have begun to go mad.

He learned that the young boy and I had fallen in love. He began to follow Artaud, through the slum's streets which now reeked of more and more revolutionaries, into alleyways which were blind. In one of them, he shot the young poet and left him for dead.

There were too many dead bodies, in those days, for there to be such a thing as murder.

When I heard this, I no longer cared what happened to W. I departed from the whorehouse. For me, there were no more men left in the world.

I had been searching for my father, in a dream, and found a young and insane boy who was then killed.

Now I stood on the edge of a new world.

FOUR

Class and Its Close Relations: Identities among Women, Servants, and Machines

Alexandra Chasin

> You see, we humans have spent centuries trying to invent ways to make all we have to do go by easier. It probably started with the wheel. And fire. Later there were tools. And servants, of course. Then computers.—Tomashoff

The author of this quote makes an implicit distinction between "we humans" and the things we have invented—things that have historically made, or will someday make, "our" lives easier. *We* live, *we* have a hard time, *we* invent things. But what about *them*, the tools? What follows from the implication that there is some class of thing that could relieve those of us who are human—by contrast to those objects, we subjects—of "all we have to do"? What follows from the history and pervasiveness of the belief that some kind of thing will liberate us from the perpetual activity of labor, that activity through which we have traditionally identified ourselves? Is it the purpose—the being or doing—of those materials that are acted upon to enable us to cease laboring? And what about the persistent assumption that there is a clear and stable ontological difference between us and them?

The servant in the list above is troubling. That servant troubles the distinction between we-human-subjects-inventors with a lot to do (on the one hand) and them-object-things that make it easier for us (on the other). Is the servant one of us or one of them, human or thing, subject or object? Or, does the servant have the kind of body that points past, or ambivalates between, the poles of this binary scheme? And what about the last element in the series above, the identity that points past the pointing past of the servant, and adds ambiguity to his/her/its ambivalence?

What about the computer? If the subject-object opposition has always been an inadequate model for understanding social identities and relations, there have always been bodies that exposed its inadequacy. In the U.S. today, computers—specifically in their tendency to engage in work, and in their imagined capacity to "replace" human workers—are the latest sort of body to give the lie to the old binary. Because identities derive from doing, rather than from being, work serves a definitive role in the distinctions between humans and our Others. Contemporary bodies that trouble those distinctions do so in, and through, labor; "all we have to do," or labor, may be the activities in and through which such bodies identify themselves as posthuman.

The "trouble" is the denaturalization of the conceptual basis for distinguishing between Subjecthood and Objecthood. These categories are, of course, culturally constructed, rather than given by God or nature, but also, their construction has historically empowered the individuals and groups and kinds that inhabit the position of the subject. "Subject" here borrows the meanings given it within the modern Western philosophical tradition that comprehends Descartes and Althusser. "Object" refers to the things that have been designated, within the same tradition, as static and stationary, passive and inert. Can posthuman bodies do more than embody an epistemological crisis, more than point to the internal contradictions of the binary schema through which we dis-identify with them? Can they help suggest an alternative to this Western-traditional model?

Other familiar dichotomies align with the opposition between subject and object: male and female, masculine and feminine, white and non-white, rich and poor, normal and deviant, mental labor and manual labor, developed and underdeveloped, strong and weak, active and passive. A great deal of work, especially recent feminist work, has gone into the rejection and/or dismantling of such distinctions. One way into this problematic is to ask how objects are not passive, or how they act socially. How do objects participate in social negotiations, in the evaluation and constant transvaluation of the categories—such as gender, race, class, sexual orientation, and nationality—that are implicated in human identities in the contemporary U.S., and in the practices that produce, re-produce, and deconstruct those categories? In particular, how do electronic machines do so?

My first consideration of the deconstructive properties of electronics followed from an exchange with my Automated Teller Machine. I noticed years ago, that when we had completed our transaction, and the machine spat my card back out, the terminal screen displayed the following message: "Thank you, Alexandra Chasin, it was a pleasure serving you." I was shocked; I wanted to ask the thing, "Pleasure?! What do you know about pleasure?" My first thought was that the reference to pleasure constituted the machine's claim to have human experience, a claim that it already implicitly makes by "speaking" in natural language. But almost as immediately, I noticed that the machine's claim to take pleasure in serving me, placed it in a certain class position, if not in a certain gender position as well. To appear to take pleasure in serving, has been, traditionally, an ideal for women, just as it has been for workers, especially servants, and often for slaves.

There are two important points to make about the ATM. The first is that the machine, to the extent that it represents itself as human, could not help but represent itself as a classed, gendered, speaker of standard American English (the official national language of the U.S.); this inevitable marking of a representation of anything human then points to the implausibility of a universal human identity, that imagined identity that has underwritten liberal humanism. In other words, the performance of humanness entails the activation of such identity markers as race, class, gender, and nationality, at least. (This entailment, in turn, suggests that such features form a weak foundation for "identity.") The second point is that the machine, in making its claim to pleasure in serving, effaces the alienation that so often attends labor, just as in its very operation it effaces the real human labor that went into its performance of service—from bank personnel to software programmers to the third-world workers who so often make the chips (O'Connor, 249).

In disabling the myth of a universal human identity, electronics confound the boundary between human and machines. Therefore, the electronics also contribute to the negotiation and renegotiation of yet another binary that aligns with the subject/object binary; that is the opposition between human and non-human, or more specifically, human and machine. Electronics participate in those negotiations partly by exceeding the definitional norms of both categories; such excesses then effect changes in the construction of the categories themselves.

Electronic machines emphatically deny the distinction between materiality and discourse, if for no other reason than that the materiality of electronic machines is so elusive; electronic devices seem to be nothing but representations. It's as though there's no there there. To identify the metal and plastic boxes, in which logic boards are so often housed, as the things-in-question misses the point, even though the most common practical interactions with electronic devices consist in human manipulations of plastic and metal. On the other hand, to identify an electron as material is somewhat unsatisfying since it is a very small particle, rarely at rest, and characterized mainly by its negative electrical charge. Moreover, when using electronic devices, most people have no direct experience of the electrons that make the devices work the way they do; that is, they don't feel like they're manipulating electrons. The distance between people and electrons in such an interaction is mediated by a series of symbolic representations—codes and languages. Through the binary code, and its translation into "natural language," electronic machines represent themselves. They may, in fact, be little other than representations of themselves; nevertheless, the fact that the metal and plastic boxes take up space nominates electronic devices as objects. If all objects contribute to negotiations of social relations, each one does so quite differently; it follows, then, that there will be very particular ways in which the electronics participate in the negotiation of gender, race, nationality, and class. Electronics also, and perhaps uniquely, contribute to the renegotiation of the boundary between humans and machines at the same time as, if not through, a reorganization of kinds of labor; that is, shifts in identity correlate with shifts in the character of certain kinds of labor. (I will return to this point.)

If the identities of humans and machines do not derive from essential features, from what can they derive? More specifically, if humans and machines both appear as endowed with identity markings that derive from their activities and interests and that reinscribe the axiomatic divisions of Western-traditional hierarchical binarisms (i.e., those that align with subject and object), then humans and machines seem to share the cultural condition of exhibiting culturally contingent identities. What, then, if anything, distinguishes humans from machines?

The traditional answers to this question involve identifying characteristics that distinguish humans from all non-humans, including animals; those characteristics are: thinking, talking, feeling, and otherwise perceiving, intentionality, and the capacity for toolmaking.[1] Based on, and reinforcing, an evolutionary model, arguments for the distinctive traits of a human species may, in one sense, place us on a continuum with animals, but in another sense, they fix our difference from, and superiority over, animals. The same ambivalence divides attitudes about the difference between people and machines. For example, Bruce Mazlish asserts that "man and the machines he creates are continuous," and proposes that we abolish the idea of "discontinuity between man and machine" (3).[2] It could be said that proponents of artificial intelligence depend on Mazlish's premise. On the other hand, the literature on electronics abounds with humanistic insistences that people and machines belong fundamentally and unalterably to distinct ontological categories. For example, in an argument against anthropomorphism in the design of user-interface systems (that is, what computers say to their users), Dr. Ben Shneiderman writes that "it is important for children to have a sense of their own humanity. They need to know that they are different from computers" (9). (I will elaborate further on these two basic positions later.) But if the question of difference between human being and machine being immediately arises with the issue of electronic activity, so, as immediately, does the question of whether people and machines act differently.

While thinking, talking, feeling, and otherwise perceiving, intentionality, and the capacity for toolmaking, are often listed as human capacities, they also, obviously, name and/or imply realms of activity. The identity or difference between machines and people, then, may be analyzed by comparing their activities at least as fruitfully as by comparing their traits. The very activity that defines machines is the activity that confounds the distinction between them and people; that activity is work. From all points on a political spectrum, people have asserted not only that machines work, but that work is their raison d'être, and quite properly so. In his introduction to the Time-Life book *Machines*, Henry Ford II writes of his grandfather, Henry Ford, "His boyhood on a 19th Century farm convinced him that men and horses were doing a lot of hard work that could and should be done by machines" (O'Brien). He goes on to quote the original Ford: ["We] have taken the heavy labor from man's back and placed it on the broad back of the machine" (O'Brien). Extending this vision toward a totalized and (this is crucial) socialized arrangement, Lewis Mumford nevertheless concurs that "That is, in fact, the ideal goal of a completely mechanized and automatized system of power production: the elimination of work; the universal achievement of leisure" (279). This essay will examine in some depth the identity and difference between the work that people do and the work that machines do. In the interest of this examination, a focus on *service* has certain advantages. First, it highlights the social meanings of work, or the relation between various definitions of work and various social arrangements. Second, it counters the marxist tendency to understand work as commodity production. Finally, electronics, as distinct from a more polymorphous category of machine, and especially as a U.S. phenomenon, suggest a reconsideration of the particular kind of labor called service—its nature and its role among social practices.

What is service and how is it related to other kinds of labor? It is a commonplace within marxist traditions to distinguish between productive labor and reproductive labor; in both cases, the telos of labor is profit. Where the former is invested in commodities whose distribution and sales profit the owner of the means of production, the latter supports and maintains the activities of production. Reproductive labor conditions the support and maintenance of cultural and technical systems, systems that, in turn, condition the uninterrupted operation of production, as well as the uninterrupted control of production, whether by owner, state, or most likely, some combination of the two. Within this schema, service is designated as reproductive labor. While this schema is itself useful, it is provocative for the purposes of this project to think of service as a kind of labor that is immediately consumed or exhausted. That is, it cannot be stored, accumulated, or saved. Service is exhausted in its performance. In this sense, service itself abets the forgetting of labor. Once exchanged, service ceases to exist, and therefore ceases to store, materially, the labor that goes into it and, equally, into the exchange.

I have just claimed that the ATM's speech and actions demonstrate the impossibility of representing a human identity that is not marked by gender, race, class, and nationality, and that the service performed by electronic machines requires a reexamination of the distinction between human and non-human identities. If, as I will go on to claim, the meanings of these kinds of identities and their markers change in relation to historical changes in types of labor (i.e., the increased prevalence over the course of the twentieth century of electric and then electronic machines in the reproductive labor force), then looking at the history of service can help make sense of the relation between machine identities and human identities.

The history of domestic service provides an especially good place to explore the relation between the labor of humans and that of machines. Not incidentally, domestic labor is subject to at least two kinds of forgetting: first, like other kinds of service, it is fundamentally reproductive, and second, it takes place in the household. As Roger Sanjek and Shellee Colen write in *At Work in Homes: Household Workers in Global Perspective*,

> Contemporary Western society is plagued with its own mystification—the ideological separation made between the household and the workplace. With capitalist organization of industrial production in the nineteenth century, much productive work moved outside the home to new workplaces. Most reproductive activity remained in the household, performed primarily by women. "Work" became something one did for a wage in a "workplace"; the home was no longer seen as the site of "work," and paid "housework" was regarded by employers as low status, even stigmatized work, or not *real* work at all. In addition to its ideological implications for gender, this separation interferes with our capacity to see the home as a workplace, and to conceptualize the interpenetration of production and reproduction. (4)

Encouraged by this approach to demystifying the division between home and workplace, and having observed crucial similarities between the nature of service in the household and the nature of service in low-status public sector positions, I move now from my ATM in the street into the household. The question here is how electronic machines—in any sphere—participate in the negotiation of such "status," and how low-status electronic service confounds the distinctions between human and machine identity.

According to Sanjek and Colen's characterization of paid household work, "at root in all cases is an employer-worker relationship" (3). The history of domestic service in the U.S., then, is equally a history of employer-worker relations, most often a same-gender, cross-class, and in the last hundred years, a cross-race, relation. In her essay "Ideology and Servitude," Judith Rollins counts the ways in which class, racial, and gender identities intersect in employer-worker relations in the domestic workplace. Among them,

> the hiring of a household worker also supports gender subordination. The middle- or upper-class female employer is able to purchase her freedom from the least rewarding, least prestigious aspects of her socially defined gender obligations. . . . She thus circumvents some of the most oppressive aspects of her woman's role—as defined by the patriarchy. (Rollins, 85)

The white, female, middle-class employer thus buys some social status for herself at the expense of a particular worker; moreover, the employer's middle-class status depends on the class structure within which she has the power to buy the services of the worker. And, as Rollins elaborates, through the exercise of her class privilege, the employer transfers her own gender subordination to the racially and economically subordinated worker. This relation typically

> afford[s] the employer the ego enhancement that emanates from having an "inferior" present and . . . validate[s] the employer's lifestyle, her class and racial privilege, her entire social world. Most important, the performance and relationship demanded function to provide the employer with ideological justification for the economically and racially stratified system in which she lives and from which she derives benefit. (78)

The history of domestic service in the U.S. bears out this analysis. Household workers constituted the largest single occupational group of all employed American women during the nineteenth century. Of those household workers, the vast majority were white women, and this phenomenon depended on the operation of slavery. In the earlier part of the century, the occupation was filled in greater proportions by U.S.-born white women, but as the century progressed and immigration increased, household workers were increasingly born abroad. English, Scandinavian, German, but especially Irish women comprised the population of foreign-born domestic workers. In the decades following the Civil War, as African Americans moved, in large numbers, out of the reconstructing South, and as factory, clerical, and sales work became increasingly available to white (U.S.-born and immigrant) women, African American women began to fill the ranks of the domestic work force in the North and especially in the Midwest. Large numbers of African American women also remained in the South, at work in white households. As one of these women reported about the town in which she lived, in 1912:

> More than two-thirds of the negroes of the town where I live are menial servants of one kind or another, and besides that more than two-thirds of the negro women here, whether married or single, are compelled to work for a living—as nurses, cooks, washerwomen, chambermaids, seamstresses, hucksters, janitresses, and the like. . . . Tho' today we are enjoying nominal freedom, we are literally slaves. (Katzman, 24–25)

David Katzman reports that, in 1890, 24 percent of female servants and laundresses were African American; by 1920, that figure had grown to 40 percent. The corresponding figures for native-born and foreign-born white women dropped from 44 percent to 39 percent and from 32 percent to 21 percent, respectively (62–63). And Judith Rollins reports that,

> Throughout the twentieth century, until the 1970 census, household work has been the largest occupational category for African American women. But since 1940, the percentage of employed African American women reported by the census as "private household workers" has been decreasing dramatically: in 1940, nearly 60 percent of all African American women workers were household workers; in 1970, 18 percent were; and by 1980, only 5 percent of employed African American women were still doing housework (cits). They are, of course, being replaced by women from Latin America, the Caribbean, and Asia. (76)

These demographic shifts indicate what large numbers of domestic workers have said about domestic work, which is that it is extremely undesirable work, both by virtue of its difficulty and by virtue of its low, perhaps lowest, status among types of work. Its difficulty and its status are, of course, mutually reproducing: its low status derives from the fact that the work is dirty, the hours are long, the benefits are nearly nonexistent; furthermore, periodic reform efforts notwithstanding, this kind of labor has always been wholly unprotected by government or union regulations. The low status of the work is also a carry-over from slavery, an institution that would certainly have cemented the association between subordinated labor and household work. Racism, and also xenophobia, must to some degree account for a low valuation of household work where such large proportions of people performing it are African Americans or immigrants. The fact of its low status, together with racism and xenophobia, appear to have licensed the inhumane treatment of domestic workers by their employers. As a domestic worker interviewed in the early 1960s said, "I wouldn't even mind the measly pay and all the demands and the bad working conditions, if only I were treated like a human being" (Gratz, n.p.). This treatment, more than anything else, has driven women to find alternative employment wherever possible, and of course the possibilities have differed according to the race of the worker.

In the last couple of decades, the demographics of the occupation have shifted again; once more, workers have come from the least enfranchised sectors of U.S. society. As Rollins points out, since Asian and Hispanic immigration has grown, Asians and Hispanics have moved disproportionately into domestic service positions. Their treatment has been little better than that of their predecessors. For example, in Southern California, in 1983, where Hispanics constituted 60 percent of all domestics, a white, female, employer published "Tell-a-Maid," a "28-page memo pad of key phrases and clip-out printed memos in Spanish and English for communicating with the estimated 100,000 Hispanic domestics in the Los Angeles area" (Holmes). This, courtesy of *People* magazine:

> Hispanic leaders in Los Angeles . . . denounce Tell-a-Maid (and its companion Tell-a-Gardener) as insulting, demeaning and racist. . . . Critics say the memos are dehumanizing. "I know it is a means of communication, but it eliminates that human contact and creates little robots," says Gloria Molina, whose L.A. district is heavily Hispanic. (Holmes)

The spokeswoman from 1912 speaks of being a slave, the domestic worker interviewed in the mid-1960s speaks of not being treated like a human being. Speaking in the 1980s, Congresswoman Molina equates the inhumane treatment of domestic workers with the creation of robots. While many factors no doubt influence the shift from "slave" to non-specific "non-human" to "robot," it may also relate to concurrent technological changes. It also brings us back, squarely, to the place of electronics.

The fitful changes—which have most often been declines—in the servant population have led many middle-class white women, in the last century, to complain of a "servant problem." The problem, for them, has been both the unavailability and the poor quality of domestic workers. This phrase was coined at least as early as 1869, when Catherine and Harriet Beecher wrote about it (NCHE, 68). The decline that has occasioned this perpetual lament coincides, historically, with increased prevalence of electric and electronic machines in the household.

Wherever and whenever the leisure class, under capitalism, grows, its growth obviously depends on structural inequalities that produce, and are reproduced, through paid (and also unpaid) household work. Such growth may also depend on the deployment of machines as household workers, which is to say that it depends on a service class of being. In other words, the social inequalities that predicate the expansion of the human middle-class in this country may depend on the ever-increasing utilization of machines as performers of work of low social status.

Perhaps since the so-called "industrial revolution," it has been inadequate to define productive labor as the work that a human laborer performs; because machines appear to perform work that is indistinguishable from human work, it is impossible to assume that the labor congealed in a commodity is exclusively human labor.[3] By the same reasoning, it is no longer adequate to define service, as a category of labor, as that which is provided by a human. If however, it is still possible to define commodities by their congealment of some kind of labor, it is possible to describe service as the kind of labor that resists congealment altogether. It is also inadequate to define a kind of labor according to the tasks or motions involved in its execution; for example, repetitive motions may characterize either productive or reproductive labor. Nor does the social class of the person performing the service (in those cases where it is a person) determine whether labor is service; for example, members of the managerial and professional classes—doctors, lawyers, bankers, teachers, bureaucratic workers of public and private employ—supply their labor in the form of services, as do nurses and household workers. Conversely, service, per se, does not designate a kind of labor that necessarily carries with it low social status. It is equally impossible to define a class of labor on the basis of whether the employer is public or private institution or individual.

Where and how do machines occupy these meanings? Mumford and Ford both suggest that machines can, do, and should work, and that their ideal role is to relieve people of labor. Referring to a traditional Western taxonomy of labor, Mumford goes on to write, "But work in the form of unwilling drudgery or of that sedimentary routine which . . . the Athenians so properly despised—work in these degrading forms is the true province of machines" (279). Again, the degrading forms of which Mumford writes are found in productive labor forms as well as service, but where service fits the description of "unwilling drudgery" or "sedimentary routine," machines may be as easily inducted into the performance of labor as they are into the performance of strictly productive labor. I will pursue the kind of service that fits that description, asking how electronics participate in the negotiation of the meanings of such work, as well as the negotiation of the identities of the beings who perform it.

Returning once more to the ATMs assists this pursuit. Having informally observed the changes in the "speech" of ATMs over the past two decades, Dr. Ben Shneiderman has described the following trend in the development of user-interface systems. In their early forms, according to Shneiderman, ATMs "had names such as Tillie the Teller or Harvey Wallbanker and were programmed with phrases such as 'How can I help you?' " ("Beyond," 6). This phase, which he designates as a moment of "anthropomorphic fantasy," personifies the machine by giving it a human (I would say American-English) name, by referring in that name to the human that used to perform (and, in another location, still performs) the transaction which the machine will perform electronically, by printing (in reference to itself) the first-person singular pronoun, and by offering, explicitly, to act like a human, that is, "to help."

If the first and anthropomorphic wave of ATM user-interface design humanizes the machines in order to attract and initiate a skeptical public, the second phase counters that humanization by naming the electronic medium. In this phase, still according to Shneiderman, "These deceptive images rapidly gave way to a focus on the computer technology with names such as The Electronic Teller, CompuCash, Cashmatic, or Compubank" ("Beyond," 6). In these names, the semiotic pendulum swings to the other side; they either imply or infer (or both) that the technology itself attracts, and even addicts, users.

"Over time," writes Shneiderman, "the emphasis moves towards the service provided to the user: CashFlow, Money Exchange, 24 Hour Money Machine, All-night Banker, or Money Mover" ("Beyond," 6). While the anthropomorphic software design emphasizes the intentional agency of the computer, and the second phase emphasizes the medium that makes possible the appearance of the computer's intentions, the third phase deemphasizes intentional agency altogether. These names reduce the machine to its functions and rhetorically remove the machine from its intermediary position between the user and the financial institution; although the machine may conduct the same transactions as ever, its name literally de-scribes it as a purely transparent (as opposed to intentional) agent. This third set of names refers to the actions that human bank tellers perform without referring to their—or to the machine's—humanity. In this way, the name begins to collapse the humanity of both machine and human bank teller into their common performance of labor.

Gaylon Howe, of the Wachovia Bank, confirms Shneiderman's observations.[4] In the early 1970s, he recalls, ATM marketing strategy assumed that the general public would be resistant to using the new technology. As a result, the interface design focused on the "personal" aspect of the ATM. Characterizing the era in which ATMs first emerged as a time before personal computers, and before credit cards had come into such prevalence as they enjoy twenty years later, Howe suggests that ATMs had to demonstrate some relation "to things people were comfortable with . . . [to] people." Concurrent with the wider public acceptance of ATMs, the EFT (Electronic Funds Transfer) technologies developed in such a way that individual ATMs were no longer constrained to serve only the customers of a single financial institution. As banks joined in regional, national, and then international networks with other banks in the early 1980s, features like network access and number of locations and variety of transaction capabilities figured more prominently in the marketing of ATMs. It is important to note that bank personnel—and not computer scientists or interface designers—made such marketing decisions.

It is nearly impossible to establish the veracity of the claim that these names form a chronological sequence in the emergence of ATMs (if the reader is an ATM user, she can, no doubt, think of some examples that support, and also some that contradict, those claims.) Regardless, Shneiderman's sequence tells a story of the moral development of the machines. For him, the difference between anthropomorphic and non-anthropomorphic styles in interface design is the difference between immoral and moral modes of computer-to-human speech. Proponents of non-anthropomorphic user-interface design think that computers that mimic human functions are deceptive, that they essentially mistreat the user. Winograd and Flores, theoreticians of Artificial Intelligence, advocate countering this trend through the design of "transparent interaction," or interface systems that enable the system itself to " 'disappear,' not to intercede in the guise of another 'agent' between human users and the computational system" (Friedman and Kahn, 11).

As quoted earlier, Shneiderman also believes that anthropomorphic design fails to uphold and convey the idea of difference between people and these machines. I would argue that if there is a problem with these representations, it is not that they confound machine being with human being, or blur the boundary between human and machine, nor that they deceive the human user about the nature or capabilities of the machine. The real problem is that they leave intact the notion that social relations depend—

necessarily and properly—on a service being, even on a service class of being. Perhaps commodities are not the only things that efface labor; perhaps service can, in effect, be made to efface itself qua labor. Perhaps, indeed, the conventions that have thus far made up the face of laboring machines, succeed exactly to the extent that they hide the labor that goes into the performance, much like the conventions of Western realist drama. In other words, the performance of service effaces labor in a different way than commodities do. The performance of service depends on people acting as though their service is not labor. In this way, whether a machine or a person serves a person, the service being must seem to enjoy serving; that is, she must hide her own labor.

The idea that machines can and should intervene into the servant problem in this country goes back at least as far as 1917, when electricity, or more precisely, electric appliances, began to figure in advertisements as ideal servants, as things that could perform labor without the displaying the liabilities of human subjectivities. At work in the household, machines have promised, explicitly, to save labor. However, as Ruth Schwartz Cowan, among other researchers, has shown, this kind of equipment does not actually save time for the presumptively female homemaker. Consequently, the labor-saving device names, and is named by, the very things it means to hide, that is, both labor and the falsity of its own premise. Nevertheless, the idea that, in the words of Time-Life, "The modern American household teems with a gleaming assortment of willing mechanical slaves," has been popular and potent throughout the twentieth century" (O'Brien, 166).

Contemporary representations of this idea often advertise electronic devices. For example, General Electric aired a series of three television advertisements for its major household appliances in 1985–86. This reading of GE's ads starts from the assumption that they reflect a standard advertising strategy, that is, that these ads attempt to seduce the viewer into buying the product advertised by tapping into familiar configurations of social relations; in other words, by representing ideologies and myths, so prevalent in U.S. culture that the viewer will certainly have already bought and consumed them, and by representing these ideologies and myths in association with products which the viewer, it is hoped, will buy and consume in the near future. Such myths are, of course, at work in association with the products advertised; additionally, the ads reveal more about both GE and the structure of those myths than could possibly have been intended either by the advertisers or by GE. Such moments of revelation are textual ironies, moments in which the true range of GE's operations are betrayed or in which the political import of the myths show through their ostensibly "informative" character; in these moments, the ambivalence of

HELLO.
(MUSIC AND BEEP SOUNDS THROUGHOUT)
CHINA CRYSTAL
RINSE & HOLD
ALLOW ME TO INTRODUCE MY MAGNIFICENT SELF.
I AM THE NEW GE 2800.
THE SMARTEST GE DISHWASHER EVER CREATED.
I WAS BORN IN THE MOST ADVANCED DISHWASHER FACTORY IN THE WORLD.
MY BRAIN IS A COMPUTER.
MY PERMA-TUF BODY HAS A FULL 10-YEAR WARRANTY.
I CAN CLEAN ALMOST ANYTHING YOU DISH OUT.
I CAN MAKE YOUR POTS SHINE.
I CAN MAKE YOUR CRYSTAL SPARKLE.
ENERGY SAVER DRYING
I CAN HELP YOU SAVE ENERGY.
LO
03 HRS. DELAY
I CAN BE PROGRAMMED
START
...TO START MYSELF.
I CAN EVEN WARN YOU
BLOCKED WASH ARM
...IF SOME THINGS GO WRONG.
SOIL LEVEL
OPTIONS
HEAVY EXTRA RINSE
MEDIUM
LIGHT ENERGY SAVER
HEATED DRYING
ENERGY SAVER DRYING
SANI ON-OFF
IF YOU EVER HAVE ANY QUESTIONS ABOUT ME...
CALL THE GE
800-626-2000
AVO: The GE 2800 Dishwasher. It does almost everything but clear the table.
We bring good things to life.
GENERAL ELECTRIC
SINGERS: GE. We bring good things to life.
We bring good things to life.
GENERAL ELECTRIC
(BEEP)

Figure 4.1

the machines, and possibly of their consumers, if not also their producers, is belied.

The first ad pictures a dishwasher, and here, he introduces himself (Figure 4.1). This ad relies on prevalent tropes of master/servant, or class, relations and likewise tells a Christian story, featuring most prominently moments of creation and sanctified procreation, which in turn involve traditional configurations of gender relations. The Christian configuration of subservient woman partner-to-man subtends the capitalist configuration of public producing man and domestic reproducing woman.

"Hello." The title of this 60-second spot is "Son of Beep" which resonates with both names of sequels in various popular culture forms, and also with Jesus, Son of God, the embodiment in human form, of God. In this ad, in its beginning, there was the word, and the word was "beep." And beep begets beeps. In the first frame, there is only the light—which is always the first move of prime movers like God and GE—and the beep. The light and the beep signify the electronic machine's basic language, which is binary, and the conversion from that language—the language of on or off, zero or one, and not incidentally, of object and subject—into natural language, and thus the machine can communicate with the human viewer. "Hello."

This same beep also connotes the air-broadcast technique for deleting expletives; in this way it is a signifier of absence, or of effacement. The beep is therefore the primary unit of the expression of ambivalence in that it effects those textual ironies, moments when the text itself refers to the very things it effaces, or at least to the fact of their effacement.

Right away, the dishwasher establishes an identity which is only partially, and not generically, human. He speaks, and speaks English. He is masculine ("*son* of beep"), but divine: here's his "magnificent self." The dishwasher announces his intelligence and refers once more to his creation, a reference that sustains the Christian subtext of the ad. He is born ex nihilo, of a sinless birth, machine from machine without the apparent contamination of human labor: machina ex deus. His humanness is foregrounded in his brain and his speech, which sounds like a seduction narrative, as though he is a sensitive new-age man who helps with the cleaning, a modern Mr. Right. He can make your pots shine, your crystal sparkle. Then he claims to help you save energy, by which he manifestly means electricity and/or gas and/or oil, but this is one of those moments where the machine betrays itself. It doth protest too much. Again, Cowan has demonstrated that women who perform unpaid housework full-time in their own homes did not, in the 1980s, do significantly fewer hours of housework per week than did their predecessors of the 1950s (the average number of hours per week may have dropped from 40 to 35) (208).

The dishwasher then announces that he can be programmed to start himself; he refers to the consummation of the sexual relationship he's been leading up to, making the move from creation to procreation, linking the sexual and religious narratives. At the same time that the big voice-over in the sky speaks, the machine appears lit up inside as though there were a ghost in the machine, perhaps a holy ghost; the singers add the choral touch. GE's motto, "we bring good things to life," associates the corporation with God once more, even as it makes explicit the permeability of the boundary between living and non-living entities, and between humans and machines.

The labor force is almost completely effaced in this ad. However, the combination of narratives of Christian creation and gendered seduction featured in this ad support procreation, or the reproduction of the labor force, materially and ideologically. The profitability of corporate production depends on all of these acts, or processes, of reproduction. If the human labor behind this performance of machine labor is not completely effaced because of the cameo appearance of the African American female servant figure at the GE Answer Center, it is only because she acts as the failsafe mechanism behind this machine. She is the very picture of a cyborg, in Donna Haraway's sense, hooked up, plugged in, located on the integrated circuit. The irony is that she is, in this same sense, a double of the machine, whose professed identity is also cyborgian. Against a background of beeps, spoken for by beeps, she is named by a number; when I called this number, the circuits were busy.

In the next ad, the dishwasher welcomes the electronic refrigerator to the kitchen (Figure 4.2). The dialogue between them brings to the foreground the military theme that was just hinted at in the last ad, an elaborate layering of failsafe systems. The fridge refers to his "smooth operation," to a failsafe backup system, and to the "rigorous testing procedures," that sound like boot camp, and which the dishwasher has also been through. Having shared the testing procedures clearly affords them a certain camaraderie, or fraternity. When the fridge notes that his interior is "ingeniously designed for efficient use of space," he betrays the entire range of General Electric's operations (indicated by their range of publications), operations which range from meteorological research to X-ray studies, from the inner space of the body to the outer space of outer space. This is another of those ironic moments when the machine speaks of that which it hides, in this case, the fact that apparently benign household appliances represent only a fraction of GE's total production—and that, at the time that these advertisements aired on television, the bulk of GE's production activity placed it among the top two or three defense contractors in the U.S.

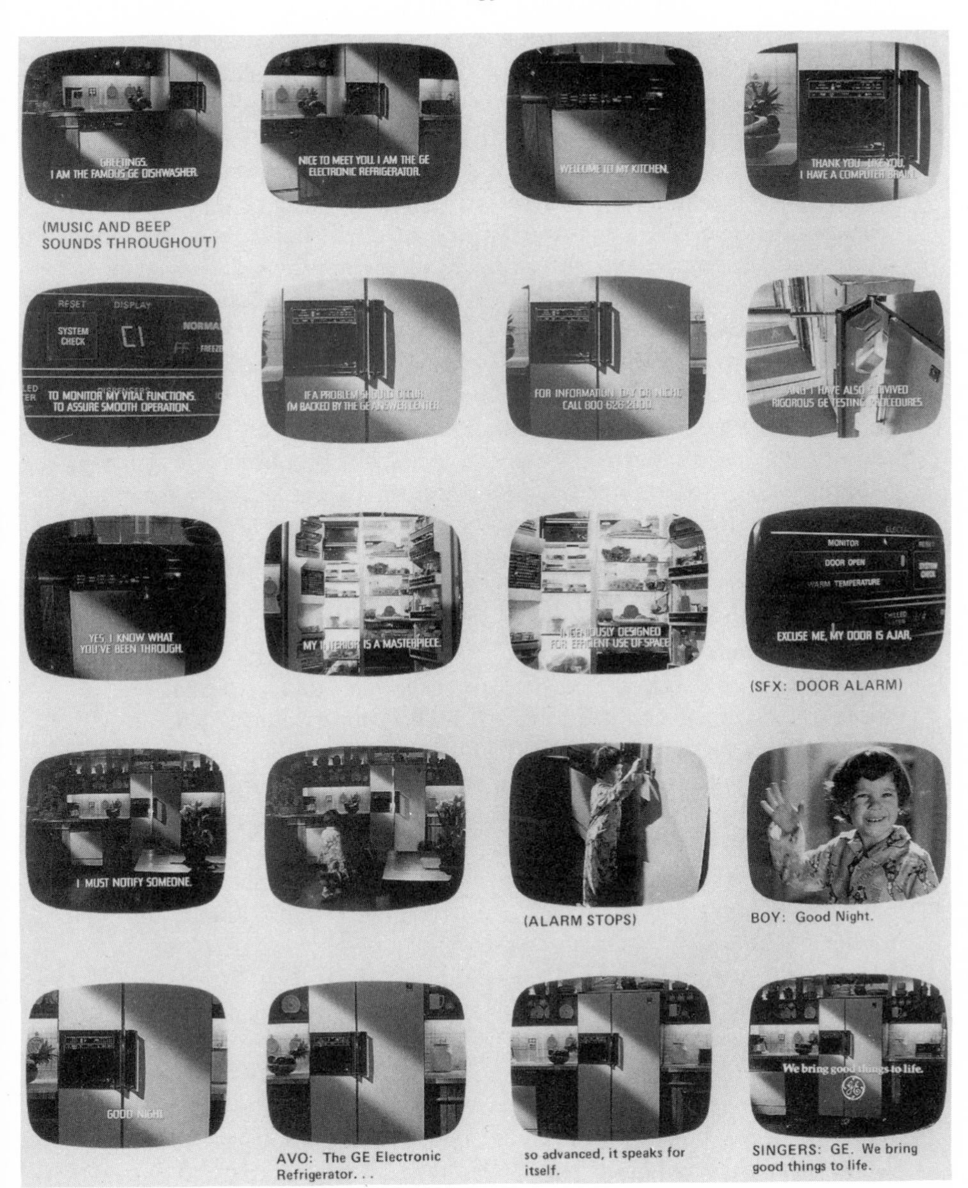
GREETINGS.
I AM THE FAMOUS GE DISHWASHER.
(MUSIC AND BEEP SOUNDS THROUGHOUT)
NICE TO MEET YOU. I AM THE GE ELECTRONIC REFRIGERATOR.
WELCOME TO MY KITCHEN.
THANK YOU.
I HAVE A COMPUTER
RESET
DISPLAY
SYSTEM CHECK
TO MONITOR MY VITAL FUNCTIONS.
TO ASSURE SMOOTH OPERATION.
IF A PROBLEM
I'M BACKED BY THE GE
FOR INFORMATION
CALL 800
RIGOROUS GE
YES, I KNOW WHAT YOU'VE BEEN THROUGH.
MY
IS A MASTERPIECE.
MONITOR
DOOR OPEN
WARM TEMPERATURE
EXCUSE ME, MY DOOR IS AJAR.
(SFX: DOOR ALARM)
I MUST NOTIFY SOMEONE.
(ALARM STOPS)
BOY: Good Night.
GOOD NIGHT
AVO: The GE Electronic Refrigerator. . .
so advanced, it speaks for itself.
We bring good things to life.
SINGERS: GE. We bring good things to life.

Figure 4.2

The assurance of the failsafe feature mimics the assurances made to the public by military experts to counter the threat of technology run amok; television viewers are assured that the machines are backed up by people, that they are ultimately under human control. The boy enters, the failsafe human in this case, and speaks to the machine. Only a child can speak with the machine-qua-servant (we will see in the next ad that the adult woman cannot speak with the machine); this behavior—unsocialized as it is—would be inappropriate in an adult. Sherry Turkle's research on children's behavior with electronic toys reveals that children speak very differently to machines than adults do. The ad implies, correctly, that an appreciation of class difference, and of ontological difference, is something that subjects mature into. The unintended military associations are made graphic in the boy's pajamas, on which the only legible word is "defense," as in Strategic Defense Initiative, Star Wars, the proposed U.S. defense plan for military control of outer space. This machine is so advanced that it speaks for itself, indeed, and also speaks for GE's involvement in the military-industrial complex, for major appliances more major than dishwashers and refrigerators.

Finally, there is the ad for the electronic cooking center, with microwave, stove range, and self-cleaning conventional oven (Figure 4.3). This ad is structured as a dramatic performance; here labor is dramatized, as are labor relations. To the audience which is constituted by the dinner guests off-screen, the woman in the ad appears to perform the labor of preparing dinner and dessert. Her performance for them involves entrances, exits, and applause. Simultaneously, the machine performs for the tv-viewing audience, both by speaking to it, and by revealing, through that speech, its conviction that *it* has really performed the labor of preparing dinner, the same labor that the woman appears to have performed for her dinner guests. If its speech humanizes the machine, then the cooking center seems to perform like a servant.

The cooking center confides to the viewer that it is the real laborer in the picture; it even makes a familiar complaint of laborers, a complaint which is of course a common-sense analysis, that the boss, or owner, is profiting from its labor, and without recompensing the laborer adequately. Moreover, it confides to the viewer its ability to speak. It keeps this ability hidden from the woman (it stops beeping when she walks in) and thus the cooking center apparently agrees to efface its labor (by not interfering with her performance-of-labor for her guests), and it also effaces its own pseudo-alienation (by not making the same complaint to her that it reveals to us).

The ad asks the viewer to identify doubly. Firstly, with the machine, whose confidence amounts to a kind of conspiracy with the viewer. If the

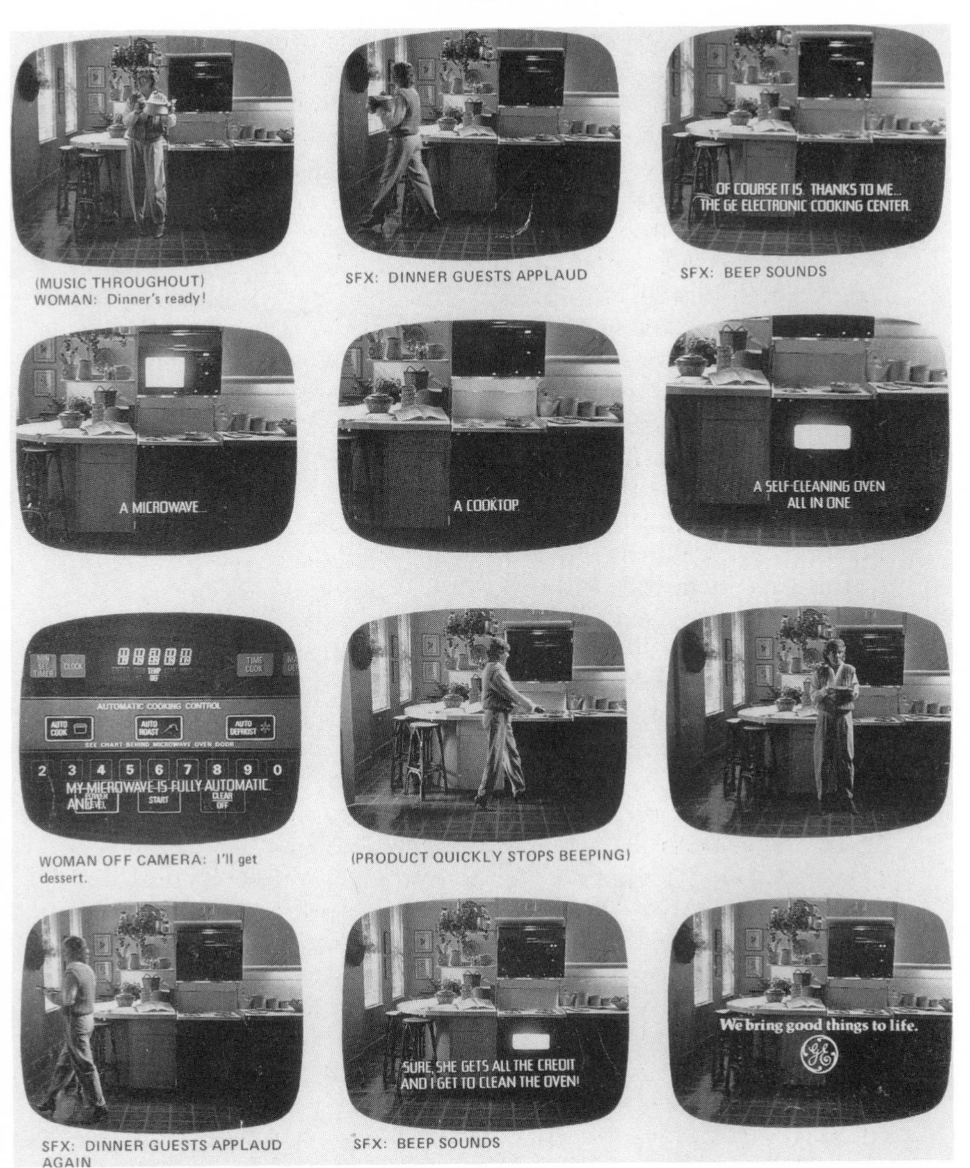
(MUSIC THROUGHOUT)
WOMAN: Dinner's ready!
SFX: DINNER GUESTS APPLAUD
OF COURSE IT IS. THANKS TO ME...
THE GE ELECTRONIC COOKING CENTER.
SFX: BEEP SOUNDS
A MICROWAVE...
A COOKTOP
A SELF-CLEANING OVEN
ALL IN ONE
MY MICROWAVE IS FULLY AUTOMATIC
AND...
WOMAN OFF CAMERA: I'll get dessert.
(PRODUCT QUICKLY STOPS BEEPING)
SFX: DINNER GUESTS APPLAUD AGAIN
SURE, SHE GETS ALL THE CREDIT
AND I GET TO CLEAN THE OVEN!
SFX: BEEP SOUNDS
We bring good things to life.

Figure 4.3

viewer understands the complaint and analysis of the machine, if the viewer accepts the machine's claim to be laboring, this is the first point of identification. Secondly, the viewer must identify with the woman, presumably by aspiring to perform for the guests as she does, as well as by aspiring to control the labor of some other entity, here, the machine. Underneath its promise that the machine saves labor, the ad promises the prospective owner of the cooking center a raise in class status, because the machine promises to perform the services of a servant. In fact, the "services" of the cooking center will not save its owner appreciable amounts of time or work.

In the frame of the woman's performance for her guests, the woman may be objectified, but the ad obscures that objectification by keeping that performance off-screen. The tv-viewing audience is not privy to the network of gazes that fix her; the tv audience is limited to the frame within which the machine, apparently willingly, objectifies itself. In effect, the subjecthood which the machine offers to the woman, it offers through its own objectification. On the other hand, the tv viewer is also made the voyeur of the woman since she is evidently unaware of the tv audience, as well as she is unaware of the communication between the machine and the tv audience. The tv audience sees her achieve a feminine ideal—a middle-class feminine ideal—when it watches her appear to her dinner guests as not having labored. The machine appears as an ideal servant by appearing (to the woman) not to have minded laboring. Put together, the woman and the machine give a performance of the ideology of labor relations, to the extent that the machine cheerily fails to represent its alienation. As in the ad in which the Answer Center woman appears, the real labor behind these machines, as well as these ads, is effaced. The same way that the Answer Center woman is obscured within the GE apparatus, the real human labor that goes into these commodities and advertisements is obscured. Here, once more the light in the oven has that remote divine quality; labor is reduced to its ghostly after-image.

The electronic appliance here enacts the performance of dominant ideology, according to which ownership and control of commodities carry with them increased social power, while it also situates the breakdown of traditional epistemological boundaries between humans and other things. These machines offer to substitute for servants exactly to the extent that they represent themselves as human-like objects—and/or objectified humans. Numerous social observers, from Aristotle to the author of the Time-Life book on *Machines*, have proclaimed the democratizing function of machines. However, to the extent that the maintenance of capitalist class relations depends on some form of subordinated labor, machines seem capable of sharing such labor with humans, the more so the more human-like they act.

Several phenomena result from this dynamic. For one thing, because the production, distribution, and use of electronics requires multiple service relations, the increased use of electronic devices may efface the phenomenon, and poor working conditions, of human service laborers, rather than obviating the existence of a service class. For another, in, and following, the era of Reagan-Bush economics, in which the middle class has actually shrunk, and the upper and lower classes have polarized, the number of electronic household appliances has risen with—rather than caused a diminishment in—the number of domestic workers.

In this climate, electronics stabilize the idea that a service class of being(s) is proper and even necessary; here, electronics participate in, and thereby reinforce, the unequal social and psychological dynamics upon which the myth of a constantly expanding middle class depends. Finally, with electronics, the boundaries between humans and non-humans are renegotiated if not altogether disintegrated. In order for class relations to change dramatically, this last proposition must be fully understood. Electronics, in general, occupy a liminal space, challenging conventional assumptions about the differences between people and machines, as well as between living and non-living entities; such challenges necessarily entail rethinking the categories themselves, the definitions behind them. It is crucial, in meeting the epistemological challenges presented by electronics, to avoid the seductive suggestion that "something must be enslaved in order that something else may win emancipation" (Winner, 21).[5] However strange it may seem, the fate of working people is inextricably linked with the fate of machines. In these times, increasingly, we work and live at the electronic-site; let us be mindful of the consequences of assuming that some being or other must serve to make it easier to live and to work.

If work is a complex of practices that ground, quintessentially, the observation that human and machine identities are hard to distinguish, it is just one set of practices that does so. Nevertheless, historical changes in the technologies of labor would concur with ontological changes in the identities of workers, as well as changes in the relations among working (id)entities. Electronics arise in, as well as inform, a historically specific—a contemporary—confusion among identities. If I have used the locution of the "posthuman" tentatively, that is because the contemporary technologies that make visible the analytical insufficiency and the political undesirability of a schematic opposition between humans and things (and hence situate its renegotiation) are merely contemporary occasions for noticing what has been true all along: "we" are an invention not less, not more, than the things "we" invent. To recur to the epigraph, perhaps the crucial invention, the one that makes "all we have to do go by easier" is not the tool-object, but is precisely the invention of difference between human subjects and other-than-human objects. More than the design of tools, the designation of absolute difference between us and them has served as a claim to entitlement, to categorical superiority, to exclusive rights to selfhood, and to selfish social and technical arrangements. Working beings have often thrown into crisis the schematic opposition between subjects and objects, just as they have objected, dramatically, to the unequal arrangements that depend on it. Here we go again.

Working machines do not replace human workers. The very idea that they could and should, lives in the heart of a humanism that preserves a separate social sphere for the human identity it also wants to preserve. However, they do replace humanisms, to the extent that they enable theories that supersede the master narratives—narratives that describe a universal individual identity as the natural, fundamental unit of the political and philosophical systems that embody liberal humanism. Not incidentally, the progressive emergence of machines that work electronically has coincided with the progressive theoretical displacement of hierarchical binarisms, and with the proliferation of theories about identity. In other words, contemporary working beings renegotiate human identity at the same time that they participate in the renegotiation of theories of identity. In this sense, beings like electronic machines situate the elaboration of a posthumanism in which masters do not designate and devalue service as the work-activity by which objects and other Others can be distinguished from those whom they rightfully serve. The theory and practice of posthumanisms will require the redesign of identities and of differences among them.

Notes

1. On the subject of intentionality, see John R. Searle, *Intentionality, an Essay in the Philosophy of Mind* (Cambridge: Cambridge University Press, 1983) and Batya Friedman and Peter H. Kahn, Jr., "Human Agency and Responsible Computing: Implications for Computer System Design," (n.p.: J. Systems Software, Elsevier Science Publishing Co., 1992). On the subject of toolmaking as a definitive feature of human being, see Sherwood Washburn, "Tools and Human Evolution," *Scientific American*, vol. 203, no. 3, September 1960.
2. Bruce Mazlish, "The Fourth Discontinuity," *Technology and Culture* 8.1 (1967): 3. Mazlish, in fact, traces a brief history of discontinuities (ideas that placed man apart from, and above, other things—ideas which had to be overcome,

and generally were when scientific breakthroughs finally proved the case for continuity). He names the men whose proofs induced "the three historic ego-smashings": Copernicus, who displaced man from the center of the universe; Darwin, who argued for our genetic "descent" from animals; and Freud, who convinced us that we weren't even at the center of our own selves, subject as we were to the workings of our unconscious. Lest this last claim seem less obvious, as a support for the argument that Freud unseated a cherished discontinuity, I quote Mazlish (who cites Jerome Bruner, "Freud and the Image of Man," *Partisan Review* 23.3 [1956]): "With Freud, according to Bruner, the following continuities were established: the continuity of organic lawfulness, so that 'accident in human affairs was no more to be brooked as "explanation" than accident in nature'; the continuity of the primitive, infantile, and archaic as co-existing with the civilized and evolved; and the continuity between mental illness and mental health" (Mazlish, 3). For Mazlish, "the combination of mathematics, experimental physics, and modern technology that created the machines that now confront us" (Mazlish, 8), confront us, precisely, with the need to abolish what he calls "the fourth discontinuity" (Mazlish, 1), the one between man and machines.

3. This statement is not intended as a proof that the work that machines do can be called labor. It is simply a suggestion that it is increasingly less accurate to define labor and kinds of labor by what kind of being has performed it.
4. In private conversation, April 1992.
5. Langdon Winner, *Autonomous Technology: Technics-out-of-Control as a Theme in Political Thought* (Cambridge: MIT Press, 1977), p. 21. Contrary to Aristotle, Giedion, O'Brien, and countless others, Winner evinces doubt that machines could fulfill the age-old promise to relieve humans of labor. This attitude toward technology, from Classical antiquity to the present, assumes that liberation is a zero-sum game; following Winner, I wonder about the consequences of that assumption and about the plausibility of the projection that technology will inevitably enable human freedom or equality.

Works Cited

Cowan, Ruth Schwartz. *More Work for Mother: The Ironies of Household Technology from the Open Hearth to the Microwave.* New York: Basic Books, 1983.

Friedman, Batya, and Peter H. Kahn, Jr. "Human Agency and Responsible Computing: Implications for Computer System Design." N.p.: J. Systems Software and Elsevier Science Publishing Co., 1992.

Gratz, Roberta. "The Career of the Last Resort." Reprinted by permission of the National Council of Jewish Women. National Committee on Household Employment, 1964.

Haraway, Donna. "A Manifesto for Cyborgs: Science, Technology, and Socialist Feminism in the 1980s." *Socialist Review* 80, 15 (2) (March-April 1985): 65–108.

Holmes, Marian Smith. "In the Money." *People*, May 2, 1983.
Katzman, David M. *Seven Days a Week: Women and Domestic Service in Industrializing America.* 1978. Urbana: University of Illinois Press, 1981.
Mazlish, Bruce. "The Fourth Discontinuity." *Technology and Culture* 8.1 (1967): 1–15.
Mumford, Lewis. *Technics and Civilization.* New York: Harcourt, Brace & World, 1934.
National Committee on Household Employment. *Report of Developmental Project: National Pilot Program on Household Employment.* (Washington: Manpower Administration, U.S. Department of Labor, Feb. 23, 1966–March 15, 1968) Feb.23, 1966–March 15, 1968.
O'Brien, Robert. *Machines.* New York: Time-Life Books, 1968.
O'Connor, David C. "Women Workers and the Changing International Division of Labor in Microelectronics." *Women, Households and the Economy*, ed. Lourdes Benería and Catharine Stimpson. New Brunswick: Rutgers University Press, 1987, 243–67.
Rollins, Judith. "Ideology and Servitude." Sanjek and Colen: 74–87.
Sanjek, Roger, and Shellee Colen, eds. *At Work in Homes: Household Workers in Global Perspective.* Washington: American Anthropological Association, 1990.
Shneiderman, Ben. "Beyond Artificial Intelligence: Overcoming 'The Obstacle of Animism.' " Proceedings of Workshop on Long Term Social Impact of Artificial Intelligence, Office of Technology Assessment, October 23, 1984.
———. "A Nonanthropomorphic Style Guide: Overcoming the Humpty Dumpty Syndrome." *The Computing Teacher* (October 1988): 9–10.
Tomashoff, Craig. "Why We Make Lists." *San Francisco Chronicle*, April 7, 1992, D3.
Turkle, Sherry. *The Second Self.* New York: Simon & Schuster, 1984.
Winner, Langdon. *Autonomous Technology: Technics-out-of-Control as a Theme in Political Thought.* Cambridge: MIT Press, 1977.

FIVE

Soft Fictions and Intimate Documents: Can Feminism Be Posthuman?

Paula Rabinowitz

Question #1: Do posthuman bodies have histories, genders, or sexualities? In posing this question the editors of this volume challenge conventional relations between the human and gender, the human and history, the human and sexuality. A simplistic reading of the posthuman might see it as beyond and before time and type, and outside the boundaries—chronological and spatial and generic—that have held humanity and humanism. While obviously the posthuman cannot claim for itself such a utopic space, still the question also fixes a certain stability onto the constructs of history and gender and sex—as if we know precisely what these highly fluid, contested and malleable forms actually are or have been before we "posted" them. It was not so long ago—as Virginia Woolf noted in her search through the British Museum's archive—that learned men could ask of women: Do you have history, gender, sex—are you human? Does that make women posthuman or prehuman? Does the term human have any meaning for women? As many feminist scholars have noted, following Foucault, the rise of the human sciences comes fast on the heels of the rise of feminine self-fashioning.[1] But in claiming space for the posthuman are we erasing yet again women's lives and stories? I am not arguing for making women human. Who needs it? Rather I want to suggest that women's stories circulate apart from human knowledge.

Question #2: Can the posthuman speak? And if so, what's there to say? When Gayatri Spivak asked her provocative question—can the subaltern speak?—she exposed the politics within posthumanist critiques of the subject.[2] Speaking is always already something done to us or for us by others whose presence as antecedents, as authorities, as interpreters, overpowers ours, even when one inhabits the most privileged of positions—that of the Western, educated, middle-class professional, like myself. How can the stories of others far outside the circulation of narrative, capital, goods, and so forth be heard? Their voices only accessible through vast networks of mediation prone to recuperation and misinterpretation at best, more likely imperial silence and violation. Poised between action and representation, posthuman bodies—voguing queens, PWAs—are bodies living outside national, sexual, economic borders. They exceed and override borders by turning bodies into acts and actions into representations. Eliminating the distinction between action and articulation, deed and word, the posthuman body is still saturated with the stories of humanity that circulate around it; it speaks through a language straddling the borders between health/sickness, male/female, real/imaginary. It tells its stories, however, through those already told; it rips off the past to refuse the future. And so the posthuman, alien and marginal like the subaltern, probably cannot speak because it is always spoken through the stories that someone else already told.

Question #3: Is there a posthuman woman? When women began "speaking bitterness" in the consciousness-raising groups of the 1960s and 1970s, women's humanity was still up for debate. Feminists sought to document women's struggle to be heard politically, historically, sexually, through the immediacy and realism of testimony in film, women's studies courses, poetry, and c-r groups. These groups, modeled on the Chinese and Vietnamese practices of criticism/self-criticism which cultivated anger and hatred within peasants and cadres where none existed—hatred being to some degree the luxury of an individualistic and mobile culture; not one based on familial, filial, ancestral ties to the land—channeled (mostly) middle-class white women's anger into political action and theory. Basic to the c-r groups was the unspoken assumption that each woman told the truth. Her story, her secret, her fury, her memory, perhaps her fantasy became the evidence from which to fashion a theory of women's oppression. But what if she were lying, or if not lying, then embroidering, weaving a fabulous story from odd encounters with the world? Feminism required sincerity for women to claim their experiences as authentically human. Perhaps a posthuman feminism develops from the evasion of truth—from fantasy, exaggeration and lies. In this essay, I want to explore this possibility (and suggest some answers to the questions asked) by examining a little-known film by California filmmaker, Chick Strand.

Chick Strand's *Soft Fiction* reveals secrets. This film signals realness and truth through various cinematic devices to allow women to voice their fantasies. However, the film's reenactments, restagings, retellings of gang rape, addiction, incest, seduction, into tales of power and control undermines and betrays the feminist-humanist project of truth-telling. As in Bette Gordon's *Variety* the fantasy and power of genre conspire to alter sexual histories. *Variety*'s heroine, who sells tickets at a 42nd Street porn theater, begins watching the films; gradually they inhabit her, but she ultimately inhabits them as she retells their plots as if recalling the day's events. Has she been made over by the images she watches, or have they provided her with new ways to speak about herself? Is she in them, or are they in her? The borders between words and deeds are permeable; acts and images dwell in the same room.

This room may be the safe space of a friend's home, where one is free to expose oneself because the thick vegetation surrounding the yard shades the interiors, because a friend will never betray a secret even though a camera is fixed on you and at some point everyone will see you, hear you. When you tell a secret after all, you expect it to get around; it becomes the substance of gossip. The exchange of secrets is fundamental to friendship, but also to power—johns tell prostitutes state secrets, informants tell ethnographers local secrets—the exchange gives away something but gets something in return: Control of the story. The mundane secrets of middle-class girlhood are divulged over kitchen sinks, across telephone wires, in private female spaces. But what if the rooms are bugged, what if a camera is there, too? Are the secrets still secret? Do the stories ring true, sound real? Hardly. And yet laden with meaning as they are we still want to believe them. More so perhaps because we know them not to be secrets anymore.

From its inception, feminism has engendered radical skepticism; once the lid was blown off and culture revealed to be hopelessly male-dominated, who could take anything seriously? Even women's authentic voices. By "speaking truth to power," women called into question both truth and power. But the joke was on those sincere believers, acting like naive ethnographers in the field soaking up authentic culture, who found women's voices pure. The fact is that posthuman bodies have been around a long time. They do have histories and these histories will be found in what has been left out of the official accounts of the marginal. Posthumans always lie. Can posthuman women speak? Of course. Will they speak to us? Not likely. Orthodoxies get established very quickly, and those out of bounds are made to disappear, kept silent, even by those of us whose job it is to listen. Feminists talk, theorize, act, but in whose (human) interest?

As I perform my role as posthuman feminist film critic, I want to suggest that this process is itself cinematic, that is, it is both a spectacle and speculative. The idea of critique as cultural performance, as posthuman activity, can perhaps point a way out of the political impasses that both identity politics and psychoanalytic theory construct for feminism. A sense of the dynamic intersubjectivity of the performance of cinema—of the bodies on screen enacting conscious performances and of the bodies in the audience taking up and remaking these performances (un)consciously and collectively—might open film to posthuman acts.[3] Something different happens in a movie theater from a dream or fantasy. Other people surround you—coughing, laughing, eating, kissing—who have also traveled to the theater, paid money, and expect affective results. In short, the performance of cinema—on screen and in the audience—defies the boundaries of individuals and their psyches by recasting them into mass formations, posthuman assortments.

Chick Strand's *Soft Fiction* recirculates many of the clichés about women's erotic and sexual fantasies within its visual and sound tracks. Bringing to focus questions about the range of female sexuality and fantasy, the modes of female address, the genre(s) of women's stories, the form of the female body as visual spectacle and narrative subject, *Soft Fiction* dwells between the borders of ethnography, documentary, pornography, avant-garde and feminist counter-cinema.[4] Its title evocative of soft core, true romance, hard fact, Strand herself describes her beautifully shot, black-and-white film as an "ethnography of women."[5] In doing so, she places it directly within the realm of anthropological filmmaking, where she began her career. Strand invokes, yet resists, the "exotic" cultural Other that forms the subject of much ethnography.

One of Strand's earliest films, *Mosori Monika* (1971), investigates the impact of a Spanish Mission on the Warao Indians of central Venezuela through the differing narratives of a young Spanish nun and an elderly Warao woman. This film exposes the missionary project as an essentially imperialist one that teaches the Indians "how to live a human life. . . . The life of a man," while demonstrating that the Warao cagily employ a form of resistance to the colonial presence of the Mission despite their apparent willingness to be clothed, fed, and feted by the nuns. Strand's narrow focus—on the stories of these two women—and her evocative close-ups of the bodies of men and women working, resting, eating, playing, break many of the conventions of the anthropological documentary by refusing to present the "whole" picture of the body or the culture. In her justification of that film, "Notes on Ethnographic Film by a Film Artist," she challenges the conventions of "wholeness" which Karl Heider had established as the mark of well-wrought ethnographic cinema.[6] Arguing for the use of extreme close-ups, fragmented movements and the "small talk" of daily life, she seeks to "get a microscopic view of one of the threads that makes up the tapestry of the whole culture." Locating the partial and the conditional, her films "evolve in the field" into "works of art" rather than scientific "textbooks."[7]

Since the mid-1960s Strand has been filming the life story of her friend Anselmo, an Indian from Northern Mexico who makes his living as a street musician. Because each film involves a level of "performance" that is self-consciously rendered to alter the "purity" of ethnographic forms, she describes the films variously as "experimental documentary," "expressive documentary," or "intimate documentary." Her first film, *Anselmo* (1967) is a "symbolic reenactment of a real event" in which Strand tried to fulfill Anselmo's wish for a "double E flat tuba." She failed to find one, but managed to locate a brass wrap-around tuba which she smuggled into Mexico and presented to him. Later, they reproduced the transaction for the camera. In *Cosas de mi Vida* (1976), Strand traces ten years in Anselmo's life as he struggles to endure poverty. The film is narrated by Anselmo in English although he does not speak the language. Strand translated the Spanish narration Anselmo provided and then taught him how to say it for the film. Again, Anselmo "performs" himself as a subject for these (in-authentic) ethnographies. Her most celebrated film, *Mujeres de Milfuegos* (1976), presents a "fake" ethnography about the "women who wear black from the age of 15 and spend their entire lives giving birth, preparing food and tending to household and farm responsibilities," by depicting "their daily repetitive tasks as a form of obsessive ritual."[8]

The idea of transforming the ethnographic film from an observational tool, one which records daily life and/or ritual as data, into an expressive, intimate, experimental documentary requires a sense of cinematic address as performative. It also presumes that cultural identities and ideas of the individual subject are constructed as performances—for the self, for others, for the camera—within various cultural arenas. Performing everyday activities as rituals for the camera undermines the concept of ritual as well as the concept of cinema. It suggests that the images on the screen respond to the capacities of the "actors" to take up one position, leave it and take up another in a stylized fashion. Perhaps this same sense of mobility—of moving in and out of a performance—occurs as well for the spectator, who, rather than being locked into a unified, or even split gaze, is always calling up various performative aspects of identity which echo, refuse, confront or merge with the screen performances. To give another example, toward the end of Trinh T. Minh-ha's film about Senegalese women, *Reassemblage* (1982), the filmmaker remarks:

> I come with the idea that I would seize the unusual by catching the person unawares. There are better ways to steal. With the other's consent. After seeing me laboring with the camera, women invite me

> to their place and ask me to film them. . . . What I see is life looking at me / I am looking through a circle in a circle of looks / 115° Fahrenheit. I put on a hat while laughter bursts behind me. I haven't seen any woman wearing a hat.

The filmmaker takes up her position among the women she has been filming and responds to their responses to her. They are happy to be filmed but the filming must be by invitation. The village women engage the filmmaking process as an exchange of looks, as a gesture and recognition of differences and so perform their daily lives for Trinh who herself becomes a spectacle for her subjects.

Recent feminist and gay studies theories of performance stress the constructedness and historically contingent nature of gender and sexual identities. Judith Butler, for instance, argues that "gender is an identity tenuously constituted in time, instituted on exterior space through a *stylized repetition of acts*."[9] Denise Riley suggests that the condition of "women" is as contested and historically indeterminant as the mythical category of "woman." Neither term—women, woman—she declares, can sufficiently pin down the multivalent claims and strategies (as Butler calls them) entailed in constituting a subject. I believe feminist politics must take its cue from queer and AIDS politics and become self-conscious of its contingent aspect—of itself as a performance responding to its own cultural space and historical period.

Likewise, I would argue that feminist film theory needs to embrace its performative quality—both as it speaks of an object productive of and produced by performances, and as it becomes another form of cinematic performance. The over-valuation of the psychoanalytic model, which reads the effects of the cinematic apparatus through the subject's unconscious responses to the imaginary, forgets just how constructed and how performative even that primal scene is. After all, the "scene" to which cinematic voyeurism supposedly refers is rarely seen—it is an imaginative reconstruction, a symbolic performance, of desire. For all its critique of narrative realism as oppressive and of critical reflectionism as vulgar, feminist psycho-semiotic film theory has perhaps unconsciously reinstated a reflectionist aesthetics by declaring the cinematic apparatus to be a map of the unconscious. Rather than describe films' contents as reflective of sociological formations (like gender roles), cinesemiotics represents cinematic form as a mirror of imaginary constructions (like sexual difference). By prying film away from its historical references—to performers, filmmakers, critics, and audiences—psychoanalytic feminist film theorists veer close to the analyses of cultural feminists.

Strand's manufacture of "ritual" performances, her reconstructions of real events, her rephrased translations, indicate to me that the ethnographic scene, as much as the psychoanalytic, is also a performance that depends on the filmmaker's desire for the encounter and the informant's willingness to act it out for an audience.[10] Films like hers and Trinh's depict the historicity of cinematic engagement by calling attention to the performances of documentary film's subjects and objects. Strand describes a recent film, *Artificial Paradise* (1986), as an "Aztec romance and the dream of love. The anthropologist's most human desire, the ultimate contact with the informant." A romance *and* a dream: a cultural construction and a psychic reenactment. Thus, as an "ethnography of women"—which is just a group of Strand's friends who visit her home in Tujunga Canyon between 1976 and 1979 to tell their stories—Strand's film challenges the notions of the "exotic" and the "whole" and of the "informant" and the "scientist" but also insists on social relations of cinematic address. "The erotic content and style" of *Soft Fiction* suggests the malleability and the pornographic, i.e., mythic (in Angela Carter's sense), quality of all fictions, including ethnographies, and the fictionality of all oral and visual testimonies, especially those of feminine desire.[11]

The film begins with a sequence of train sounds and horizontally moving patterns of light and dark. It takes a few minutes to orient oneself to the sound and image which finally resolve into a close-up of a woman's face against a window. She departs the train and like Maya Deren in *Meshes of the Afternoon* (1943), to which *Soft Fiction* pays homage, walks in the late afternoon Southern California sun through some greenery to a locked house. She remains outside, but the camera enters the room and surveys it voyeuristically: checking the kitchen, bathroom, bedroom—recreating the dizzying descent of the staircase in *Meshes*—grazing the shadow of the filmmaker herself to discover a woman sitting in an armchair near a window calmly smoking as a woman's voice-over exhorts us to "move, first one way then the other—gathering, lifting, squeezing, releasing, just so it feels good." This reference to counter-cultural California sensuality also nods to a female avant-garde film history.

Beginning with Germaine Dulac's *La Souriante Mme. Beudet* (1922) and continuing through Deren's *Meshes* and Menken's *Glimpses in a Garden*, much women's avant-garde cinema develops as an exploration and exposure of interiors. Deren's and Strand's cameras scrutinize the empty houses they enter, but these houses are their own, turning the voyeuristic gaze into an exhibitionist display of its objects. However, where Deren multiplied her own body to display the terrors and desires of the female subject, Strand includes the voices of many women to demonstrate the multiplicity and resistance of women's fantasies. Like Carmelita, the Warao woman, whose incorporation into the mission can be read as victimization but whose own rendering of it challenges us to read resistance in her very acceptance of the nun's offerings, the women tell stories of incest, addiction and Nazis which are potent tropes for women's victimization. Yet the women's voices, the images they construct to accompany their tales and the sequencing of these images counter preconceptions of female powerlessness by substituting in its place the power of acting. The challenge to politically correct feminism has a forgotten history. Long before Camille Paglia and Katie Roiphe were condemning feminism's embrace of female victimization, Strand and her "informants" were exploring, even celebrating, their politically incorrect desires, fantasies and experiences.

As the seated woman begins her story, initiated by rubbing the curving banister of the Pasadena Art Museum, her desire to "become this railing—become this piece," invites us to question the very terms of representation that objectify women's bodies. That the play on the word "piece" is deliberate, we hear in her slow, precise language. We see her lips, nose and eyes peering directly into the camera; cinematic convention tells us she is revealing truth. The camera leaves her as she asks, "Haven't you ever wanted to live within black fur?" The tactile transvestism of this woman's desire—to inhabit curved alloyed metals, black fur, to turn her body into an object of touch—destroys the sensation of inside/outside for us as it extends the body into new spaces, new desires. It also transgresses both cultural feminism's and psychoanalytic feminism's rigid resistance to (yet ironic insistence upon) woman's objectification. The speaking subject of this sequence desires objecthood.

Another woman appears intently studying a piece of paper with a magnifying glass before she begins to read a letter addressed to Strand recounting the story of a photographer whose escapade at a rodeo she had gone to shoot ends with her giving a series of blow jobs to anonymous cowboys in a dark dormitory room. The incidents seem "inevitable" to the letter writer; her loss of control at the rodeo becomes visible to the letter reader in her handwriting—she fails to capitalize her "I"s. Already mediated on several levels (the woman's story appears as a letter written to Chick but read by a giggling woman through a magnifying glass to the camera), her story is deeply ambivalent. Has she been coerced? Is this a case of gang rape? Or is it a staging of a fantasy which oscillates between her power—as voyeur, as photographer—and theirs—as exhibitionists, as sexual cowboys? After she and her camera escape unharmed from this encounter, she picks up yet another cowboy to shoot. He takes her to a stable where she photographs him naked except for his belt, hat and boots—the regalia of s/m scenes—and where again she gives him a blow job while his buddies watch. Her fear is countered by her excitement, which is mediated further by his final remarks of comfort: "It will make a good story to tell your grandchildren." In a bizarre re-ordering of the female oral tradition, sexual pleasure exists for the man in his fellated orgasm, but for the woman, who never quite gets off herself, it is deferred, available only in the verbal recreation of desire through memory, narrative, and performance.[12]

In this story, Strand and her informant manipulate one of the privileged scenes of hard-core porn, the blow job, evoking visceral reactions from audience members about the woman's status as a "victim."[13] Linda Williams has argued that the growing popularity during the 1970s of feature-length porn films, such as *Deep Throat, The Devil in Miss Jones* (Damiano, 1972), *Behind the Green Door* (Mitchell, 1972), signaled that there was an audience for the visible evidence of desire as a fetishized commodity and that mass media could produce it. These films invoked women's demands for more and better sex through fantasies that fulfilled male desire, thus resisting the threat of feminism by constructing women's desire as a turn-on for men. Like the radical feminist Anne Koedt, *Deep Throat* rejects "the myth of the vaginal orgasm," but, as it orchestrates its "sexual numbers" around the ejaculation of fellated penises into Linda Lovelace's ecstatic face, its ultimate audience is male.[14] Still, the narrative appeal to a broader audience (one that presumably included heterosexual women) refracted the messages of soft-core melodramas, such as *Looking for Mr. Goodbar* (Brooks, 1977), which also assumed women's independent desires for sexual adventure, but provided cautionary tales about the dangers of arousing male sexual aggression for their largely female audiences.[15] The soft-core films looked back to the 1940s woman's film genre, and to the popular woman's romances found in *True Confessions*, where transgressive sexuality in a woman always resulted in shame and punishment.[16]

But in *Soft Fiction*, the photographer returns to her pleasure and her power. In her ironic reply to us, not to her handsome cowboy, she asserts, "Well, photography is a power to be reckoned with," revealing that after she prints his photo she discovers his name on his belt, tracks him down, obsessively follows him home, and declares "I know where he lives now." As Strand says of all her "informants," they take "responsibility for having had the experience. It's not that they take responsibility for the experience happening but for 'having had' it."[17] The claim of "responsibility" challenges women's victimization in/by narrative by asserting that their stories are conscious re-enactments. The process occurs as a translation—a re-fashioning of the experience into a narrative and visual sequence. This recurs at various times throughout the film which continues to switch codes between the expressionistic frame of the woman's quest (for pleasure?) and the concrete documentary-like stories women tell about the real and fantasized causes and effects of these quests—stories of pain, violation, and desire.

The next shot reveals a sun-drenched kitchen. We watch a nude woman enter, her body strong; she is unself-conscious of it as she flips on the radio before she starts preparing a hearty breakfast of juice, coffee, eggs, buttered toast. The show, *Grand Central Station*, begins with the sounds of a train over which the narrator intones "this is a love story," reminding us that "the door to the great white way is usually through the back alley." At this point, we hear a voice-over as a woman describes a sensuous memory of swimming in a pond as a child—diving "in and up and down"—until tired she ran to her grandfather waiting with a towel. She describes walking back to their cabin watching the drops of water splashing in the dust. Then, matter-of-factly, states, "I was young, only seven. We would make love on the couch, the red couch, I trusted grampa—even fell into enjoying it." She describes how he kissed her and undressed her, noting that only once did she see his penis . . . "like a snake, a pink velvet snake . . . he used it on my clitoris . . . he wanted to teach me how to make love . . . how to be sensual." By this time, she is eating. The camera no longer displays her whole body, but again is extremely close-up. Cutting into the egg yolk with the side of her spoon and smearing it over the whites, she remembers how "it scared me—it was too close and too strong. . . . I just wouldn't allow myself to be alone with him—jump out of bed, feign sleep, all the typical tactics of female avoidance—I learned them young—now I'm a master. Pursued and captured, really captured cause there's no way out," she declares as she exhales her after-breakfast cigarette.

Again, ambivalence is crucial to the performance of this scene. Hers is the only body we see whole, performing a whole act, her story distanced by the off-screen narrator. Her voice is strong, ironical, yet vulnerable. She is angry, but the circumstances she has constructed to disclose her secret imply that she has power over them. Hers is certainly "Not a Love Story" and the "responsibility" is certainly not hers, yet her image and her story—its disembodied narration running over her real time act—unpacks the cinematic baggage this story of female powerlessness holds for us and perhaps her.

The film now cuts to a clichéd image of feminine eroticism as a nude woman dances to Sidney Bechet's rendition of "Petit Fleur," and we see the play of light and dark as her body and hair break the sun's rays. This diversion momentarily breaks the tension of captivity encoded in the woman-in-the-kitchen's story and in the previous use of the extreme close-up in the woman-at-the-museum's and rodeo's stories. But the next woman tells about being "really hooked." Again, lips, eyes, brow are prominent in extreme close-up, as the woman chain-smokes, drinks wine and describes her "plan," her "program," to become hooked first on a man, then on the pain he caused, then on heroin to escape the pain. Ultimately, she kicks, despite wondering why: "It was so good, so clear, so real, so spacious. But I did it—that was the plan and I exorcized him." The exorcism extends into the filmmaking process. Strand claims that "the most incredible part of making the film was my relationship to the women when they were talking and being on camera, and doing it knowing the result, knowing that they would be on this big screen and a lot of people, strangers, would see them . . . them telling it on camera acted as an exorcism. . . . "[18] In other words, the informants became self-styled performers for an audience who was both distant from, yet intimate with, the "connections" in these stories.

The complicity between storyteller, filmmaker, and audience in the production of a voyeuristic fantasy continues as we watch a dog arranging itself into a comfortable position on an armchair. The sound track is of loud voices—a train station perhaps, no, an audience at a performance who breaks into applause when the dog stands, revealing that it has only three legs. Then a white face and white hands emerge from the blackness, the woman begins singing Schubert's "Death and the Maiden," whose lyrics evoke the longing for "dreams," for "sleep." In calling forth the romantic vision of desire as death, the conventional reading of women's masochism is reinserted as a commentary on the women's stories so far. Yet, by doing so through a soprano's rendition of the *lieder*, the female body as a performative tool is reasserted. Although each story has been painful—we see their faces contort, hear their voices crack as they speak—they have all been humorous as well. Each woman has restaged her "tragedy" into a story of power and pleasure by the styles of their telling and the compositions of their images. Still, these tropes of captivity are the stuff of female masochism, their "true confessions," the stories of surrender and desire that fueled my politically incorrect preadolescence. As if to confirm our secret complicity with the mechanisms of pornographic surrender, after the *lieder*, we find the traveling woman again. Watching her depart the house, her suitcase opens exposing yards of cloth and a sequined teddy.

The final segment frames a tight, nervous face: "Okay," she says, "this is going to be a little bit difficult." Her story is set during the war in the Poland of her childhood, when she says "it was demanded of me that I stay quiet . . . people were after you." It is not clear whether her family was hiding Jews or were themselves Jews in hiding, but after a neighbor informs on them the Gestapo visits her home. She "remember[s] being put on a Gestapo officer's lap to divert his attention—I understood that—what my job was. . . . I remember flirting with him." Her faltering voice continues with a memory of being awakened by her mother when she was 3 1/2 and walking at night for miles:

> It was necessary for me to be very brave. I remember that I liked that and I remember that I like that now—that I was brave then. And I remember a hill with fire and explosions of all sorts. I remember how frightened everyone was and my father carrying the bird cage with the kittens. And I remember feeling proud that I didn't want to be carried. And I remember that hill and there was something very bad going on on the other side, and then there's a blank.

Unlike the preceding stories of sexual adventure and surrender, which emerge as coherent, well-plotted narratives—the stuff of conventional melodrama—these memories of historical necessity are fractured, disrupted, and lack clarity. Yet, even here, the sensation is of control, of the power this young girl experienced despite, or rather because of, knowing she was an object of exchange in a larger transaction.

The last images of the film return to more clichéd images from soft-core porn—a woman's abandon as a shower of clear water washes over her, a woman walking barefoot along the shore, two naked women frolicking on horseback. These also are the clichés of California independent filmmaking of the 1960s—the sensuous display of the body at play in nature. Like the Schubert *Lieder*, the train journey, or the solo dance, they recontain the stories of female transgression and pleasure in the face of masculine power within the limits of conventionalized depictions of female desire. Yet the stories undercut this containment, violating boundaries, just as the excesses of the extreme close-ups explode the documentary conventions of the talking head by overvaluing the partial elements of the face—lips, brows, nostrils—and body—hands, legs, feet. These fragmented, cut-up icons of femininity that appear commodified in advertisements have been recharged by the speakers. By allowing movement in and out of frame, the stationary camera enables the speakers to take control of and produce their images.

The stories in *Soft Fiction* flow out of each other—the way one might reveal secrets to a stranger on a train. They are intensely private and personal, yet by orchestrating them within the compositions of avant-garde cinematography, documentary address, ethnographic filmmaking and soft-core porn, Strand wants us to begin questioning how female pleasures are experienced and represented in patriarchal culture. The film's ethnographic inquiry seems to ask, what are the narrative and visual components of white middle-class women's (hetero)sexualities? How are they represented and performed in bourgeois culture? The stories acquire their meanings through a complex interplay of image, sequence, and sound. The tight framing of faces restricts women's bodies as cinematic spectacle, yet we also participate in a voyeuristic invasion of private space and conventionalized modes of displaying female desire as well as listen in on some juicy secrets. The stories seem private, yet their performances enter public spaces. In the process, the tropes of the victim are recast through the process of storytelling into a grand panorama leading from narcissistic joy to genocidal horror. As they invoke a history of genres—melodrama, case study, gossip, romance—all too familiar in their containments of women's desires and their commodifications of women's pleasure and pain, these stories ask us to step outside of conventional narratives and images put forth by hard and soft core porn, and by their anti-porn feminist critics, to allow for the possibility that the "story is a sexual fantasy lived out."[19] In their development of women's powers of performance—powers depending upon cultural contradictions that recognize both the Oedipal narrative's power over, but also its ancillary status for women—this film also challenges the psychoanalytic model of spectatorship. In so doing, the verbal and visual performances of desire present what Adena Rosmarin calls "the power of genre" as a put-on, because here the genre's power is put on and displayed through its clichés.[20]

Strand's limited ethnography provides a partial view of the culture of heterosexual practices that are both oppressive and pleasurable to women. Her picture of white, middle-class women's culture owes much to the boy-crazy girls' gossip sessions I remember from junior high school slumber parties in which secrets, fantasies, and homoerotic desires merged with popular cultural renderings of woman's surrender. However this fantasy depends upon and fuels the racism and class division that produces the fantasy of women's culture in the first place (for instance, the only black woman on screen is seen dancing nude to Bechet—jazz and the black woman's body being icons for white dreams of sexual escape). In short, the film becomes retrograde in its obsessive explorations of white, middle-class heterosexuality.[21] Thus, to a certain extent, *Soft Fiction* participates in the anti-porn feminist (and American New Right) hysteria that elides women's victimization by male sexuality with genocidal practices of fascism as it moves from the private fantasy experienced outside political contexts to the intrusion of military force into the domestic space. In addition, the straightforward presentation of women's voices, coupled with the ecstatic images of female sensuality appear as "unsophisticated" representations of desire.[22] These distortions reveal the fault lines of, because they stem from, Strand's investment in a universalized vision of women's culture.

Moreover, Strand is caught in a serious dilemma when she embraces (albeit critically) ethnographic cinema. On the one hand, ethnography as a historical practice in which white people look at and (through cinema) display people of color maintains imperialist relations of domination. On the other hand, by turning the lens on her own culture—that of white middle-class film artists—Strand's ethnography of women would seem to rectify the colonial relationship of ethnographer to informant. But by removing her lens to her own backyard, *Soft Fiction* places the third world under complete erasure. In either case, as sympathetic yet still colonizing spectator of the other, or as empathetic exhibitor of the self, Strand's films, by invoking ethnography, despite problematizing subject and object, inevitably fall victim to their own tensions.

Nevertheless, her films depend on the performances of their informants' memories appearing simultaneously authentic and constructed, human and posthuman. As restagings, the artifice involved in the deliberate diction and the claustrophobic framing of *Soft Fiction* constructs speaking subjects who eventually call into question the possibility of a "culture" of women about whom one could make an ethnography. The stories indicate the ways in which culture as a human practice may be irrelevant. Posthuman feminists perform the competing collective strategies of storytelling and acting we carry out in all their contradictory modes every day. That may be woman's culture, thereby calling forth an ethnography, but maybe it's something else instead.

I'm not seeking to rehabilitate *Soft Fiction* by inserting it into an existing canon of films. I would hope that my discussion of the film has pointed up some of the polarized positions within feminisms—anti-porn/pro-sex, cultural/psychoanalytic—that verge in their drive toward purity and truth on the aestheticization of politics that Walter Benjamin called fascism. I believe we need to rethink the categories governing our political and cultural theorizing in order to begin the posthuman project of (re)politicizing art. Second-wave feminism in the United States, like its predecessors, has relied on cultural performances—from the Miss America Pageant demonstration through the Women's Pentagon Action to the Guerilla Girls' recent billboards—to foreground politics. Theory also might best be considered as a performance—a collective playing out of cultural codes in public spaces that are socially and historically constructed and reconstructed in response to political challenges. Strand's film and the many other expressions that step out of bounds demand that we constantly inspect the ways, in the name of political correctness or theoretical sophistication, we police the borders of feminism.

Notes

1. See the work of Nancy Armstrong, Mary Poovey, Denise Riley, Anita Levy, Cathy N. Davidson, among others.
2. Gayatri Chakravorty Spivak, "Can the Subaltern Speak?" *Marxism and the Interpretation of Culture*, ed. Cary Nelson and Lawrence Grossberg (Urbana: University of Illinois Press, 1988): 271–313.
3. I should note that in "Changes: Thoughts on Myth, Narrative and Historical Experience," Mulvey reconsiders the spectacle/spectator model, situating it as a polemical intervention made at a precise moment in the political histories of feminism and film. *Visual and Other Pleasures* (Bloomington: Indiana University Press, 1989): 159–76.
4. Patricia Mellencamp describes the strategies of "heterogeneity" in recent feminist film and video: "(1) the emphasis on enunciation and address to women *as subjects* . . . (2) the telling of 'stories' rather than 'novels' . . . (3) the inextricable bricolage of personal and theoretical knowledge; (4) the performance of parody or the telling of jokes . . . (5) an implicit or explicit critique and refashioning of theories of subjectivity constructed by vision; and (6) a transgression of boundaries between private and public spaces and experiences," *Indiscretions* (Bloomington: Indiana University Press, 1990): 130-31. *Soft Fiction* employs virtually all of these strategies.
5. This was how Strand described *Soft Fiction* in a public lecture before its screening. Ann Arbor, MI, November 1979.
6. Karl Heider, *Ethnographic Film* (Austin: University of Texas Press, 1976) states: "A basic principle of ethnography is holism. . . . From this principle come the related dicta of 'whole bodies,' 'whole people,' and 'whole acts' " (7).
7. Strand, "Notes on Ethnographic Film by a Film Artist," *Wide Angle* (1978): 45–50. This is precisely the same point Trinh T. Minh-ha has made about her controversial "documentary" about Senegal, *Reassemblage* (1982). "I knew very well what I did not want," she says of making the film, "but what I wanted came with the process. . . . My approach is one which avoids any sureness of signification . . . the strategies of *Reassemblage* question the anthropological knowledge of the 'other.' " Constance Penley and Andrew Ross, "Interview with Trinh T. Minh-ha," *Camera Obscura* (Spring-Summer, 1985): 89, 93.
8. *Canyon Cinema Catalogue* 6: 221–23.
9. Judith Butler, *Gender Trouble: Feminism and the Subversion of Identity* (New York: Routlege, 1989): 140.
10. This goes back to Robert Flaherty's reenactments of the whale kill in *Nanook of the North* (1925).
11. Marsha Kinder, "*Soft Fiction*," *Film Quarterly* 33 (Spring 1980): 50.
12. Thanks to Jane Gallop for this pun which she made in response to the panel,

"En/Countering Censorship: Feminist Transgressions," at the 1990 ASA Convention in New Orleans at which I presented a version of this paper.

13. "Chick Strand at the Cinematheque," *Cinemanews* 3/4/5 (1980): 11.
14. Anne Koedt, "The Myth of the Vaginal Orgasm," *Notes from the Second Year* (New York: Radical Feminists, 1970). See Linda Williams, *Hard Core: Pleasure, Power and the "Frenzy of the Visible"* (Berkeley: University of California Press, 1989) for a full analysis of the genre of feature-length porn films.
15. See Ann Snitow, "Mass Market Romance: Pornography for Women Is Different" in *Powers of Desire*, ed. Snitow, Stansell, and Thompson (New York, Monthly Review Press, 1983): 245–63.
16. See Ann Snitow, "Mass Market Romance: Pornography for Women is Different"; Tania Modleski, *Loving with a Vengeance* (New York: Methuen, 1986), and Janice Radway, *Reading the Romance* (Chapel Hill, University of North Carolina Press, 1987).
17. "Chick Strand at the Cinematheque," *Cinemanews*: 11.
18. Ibid., 1.
19. Ibid., 14.
20. Adena Rosmarin, *The Power of Genre* (Minneapolis: University of Minnesota Press, 1987).
21. However, in the question-and-answer session after screening *Soft Fiction*, Strand outed the woman in the kitchen, saying, "She's fine, she's a lesbian now." Public lecture Ann Arbor, MI, November 1979.
22. See Kaja Silverman on the "sophisticated" feminist films of the avant-garde. *Soft Fiction* is not among them. *The Acoustic Mirror* (Bloomington: Indiana University Press, 1988): 153.

SIX

Reproducing the Posthuman Body: Ectogenetic Fetus, Surrogate Mother, Pregnant Man

Susan M. Squier

> Nature and culture are reworked; the one can no longer be the resource for appropriation or incorporation by the other.
> —Haraway, *Simians* 149

Four figures anchored the nineteenth-century preoccupation with sex as the object of expert knowledge, according to Michel Foucault: the hysterical woman, the masturbating child, the Malthusian couple, and the perverse adult. As the separation of sexuality from reproduction aspires to technical completion in the postmodern era, we can ask what figures now anchor our understanding of the other term in that copula now nearly under erasure: reproduction, now at the boundaries of the posthuman. As a way of beginning to map that terrain, this essay will trace the lineage of three images foundational to our contemporary preoccupation with reproduction as the object of expert knowledge and power: the extrauterine fetus, the surrogate mother, and the pregnant man (Foucault 105).

These three images serve as bodily sites for potentially oppressive scientific/technical interventions, yet the [re]construction of the human being they depict is not uniformly negative. By tracing some of the fantasies these images enact in the cultural imaginary, I will show that they play different roles depending upon their institutional and generic positionings. While the images frequently express a disturbing tendency toward gender-based objectification, they can also articulate more emancipatory models of the human subject, and human relations—models that displace the binary constructions of modernist epistemologies (Haraway, *Simians* 187).

Neither inherently oppressive nor inherently liberatory, this trio of images marks the point at which—with a return to "nature" no longer possible—reproductive technologies are producing the posthuman. Whether the result will be the emancipation from certain fixed and historically op-

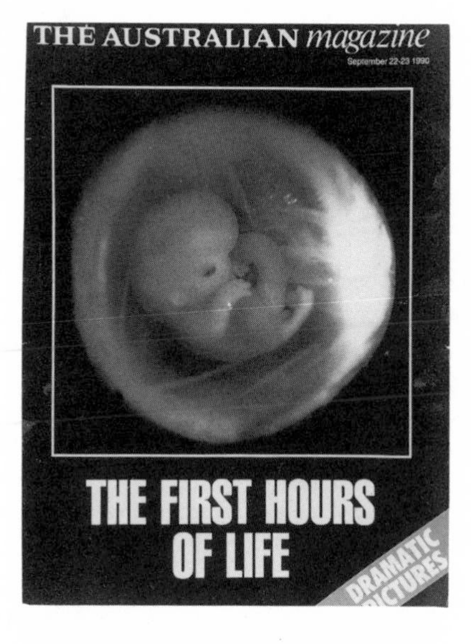

Figure 6.1. Cover of *The Australian Magazine,* September 22–23, 1990, featuring an image of an extrauterine fetus. Used with permission.

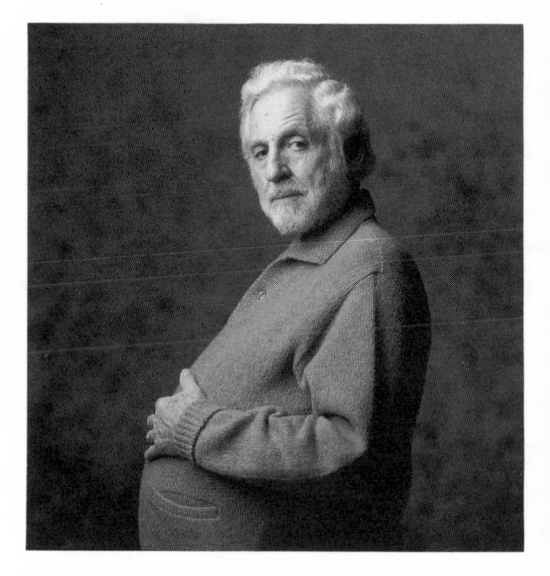

Figure 6.2. "The Father of the Pill," Dr. Carl Djerassi, as a pregnant man. © Copyright Jock McDonald 1992.

pressive constructions, the cementing of a new, even more oppressive set of social relations, or—as is most likely—both at different times, depends not on the reproductive technologies themselves, but on the social and cultural conditions of their use.

Two crucial features mark these contemporary images of reproduction.[1] First, in each image, *the fetus is separated from the gestating mother-to-be:* whether left to float freely in a bottle, in space, or inside the globe; resituated in another "mother"; or (most shockingly) placed in the body of a man. Although this separation is graphic in the representations of the fetus and the pregnant man, and only implicit in the figurations of surrogate mothering, the images have as common denominator the process of de- and recontextualization, a process whose postmodern implications are rooted in the Romantic and modern history of reproductive representations.

Second, these images all represent different versions of what we now call reproductive technology, versions ranging from the actual to the hypothetical, from the visually explicit to the visually veiled. Yet the images are not the same as the medical practices they represent. A gap exists between the range of medical practices (actual and hypothetical) known as reproductive technology, such as AID, IVF, Gamete Intrafallopian Transfer, Zygote Intrafallopian Transfer, Zona drilling, abdominal pregnancy, cloning, and so on, and their representations. As material practices that have a low success rate, the potential for iatrogenic health damage, and a problematic relationship to a pronatalist culture and society, reproductive technology has been indicted by feminist social scientists as: unsuccessful, unsafe, unkind, unnecessary, unwanted, unsisterly, unwise (Stanworth 290–91). Convincing as that critique has been, it fails to take note of two important points: that the social and cultural conditions of their use have shaped the impact of these technologies, and that representation has played a crucial role in setting the social and cultural boundary conditions for their use.

Literary figurations of the reproductive body have always been open to a wide range of meanings, because literature functions as one of the institutions through which human beings are shaped to the needs of their society, through the process of identity construction that occurs in, and mutually implicates, both the symbolic and the material realms. As Catherine Belsey has argued, representations help to construct what we understand as our cultural and social experiences, including our experiences of the body (Belsey 593, 598). I will argue that figurations of reproduction—in literary and non-literary texts—performed three functions in the Romantic and modern periods, all related to the production of power: the creation of a metaphoric break between mother and fetus that enabled their different social positionings; the reconstruction of woman's body to produce or consolidate male power; and the reconfiguration of the [male and female] human body to serve industrial production. I will trace those different ways of producing power as they operate in representations of a range of technological interventions in reproduction, from the hypothetical, to the obstetrical, to the biomedical; from the Romantic era, through the modern, to the postmodern.

While reproductive technology shifted from a hypothetical to an actual medical practice during the period under discussion, it is not the break in material practice that concerns me here, but the continuity in representational strategies. Initially reflecting the European context of Romanticism, the range of these reproductive representations expands in the modern industrial era to reflect British and American concerns, and by the postmodern moment, the images reflect and help to consolidate the global power of multinational late capitalism. These representations both converge and diverge in the postmodern era, as an examination of four contemporary texts dealing with aspects of reproductive technology—

three fictions and one non-fiction report of a government committee—will reveal. Producing different origin stories and constructions of identity, these representations function to define and distribute difference, within and across a variety of temporally and geographically overlapping power grids: civil society, psychoanalysis, industrial capitalism, institutional science, and medicine.

Mary Shelley's *Frankenstein* inaugurates the Romantic moment with its image of male birth. Written at a time when female procreative power was being co-opted metaphorically to represent the birth of fraternal contractual democracy, *Frankenstein* figures as monstrous the male monopoly on political creation. Shelley's monster, as outcast, contests the inclusive, egalitarian pose of the liberal civil state. Yet the reproductive critique articulated by Shelley's novel coincided with shifts in contemporary medical science that enabled, rather than critiqued, ongoing political changes. Around the time that *Frankenstein* was published, the fields of embryology and obstetrics authorized new representations of the pregnant woman and the fetus harmonizing with the new political arrangements. Late-eighteenth- and early-nineteenth-century embryology affirmed the theory of epigenesis (the notion that an embryo develops from lesser to greater organization in the course of gestation) over the earlier theory of preformation (the notion that the embryo is a static, preformed, miniature entity, somewhat like the homunculus). This shift in scientific knowledge joined Rousseauean notions of child-rearing to produce an individual fitting the needs of bourgeois capitalism. As Andrea Henderson observes, "Early-nineteenth-century epigenesis sketches us a picture of the Romantic fetus [as] the perfect bourgeois subject—it makes itself, and so is neither simply the inheritor of paternal power nor the commodity-like product of its mother's labor" (Henderson 112–13).

Parallel to the victory of epigenecist embryology was a shift in the representation of the gestating and childbearing woman in anatomical engravings and midwifery manuals. Emphasizing the bony structures of the maternal pelvis as objects to be manipulated by equally rigid obstetrical instruments, these illustrations articulated a "trend . . . to present childbirth as a mechanical process, having affinities with mechanical production, but with the role of the woman . . . in the productive process . . . not as laborer but only as a machine" (Henderson 103). Like *Frankenstein*, these non-literary representations participated in the Romantic [re]construction of reproductive subjects—man, woman, fetus. They reshaped the fetus as the state's ideal, organically developing, autonomous individual; they marginalized woman, exiling her from the public realm of the social contract to the private realm of the sexual contract; and they reconstructed man as both father and mother of the new political order (Pateman 88).

Modern representations of reproductive technology built on the romantic separation of developing fetus from machine-like mother to serve ends not political but productive, and ultimately industrial. Fueled by racism, misogyny, and classism, the early-twentieth-century eugenics movement aspired to the bioengineering of a racially uniform (white) and genetically "perfect" human species, fantasies that achieved culmination in the Nazi era. Yet, paradoxically, there were also eugenics enthusiasts who were motivated predominately by their belief that the biological substrate of human experience was not natural and unchanging, but constructed, and hence subject to manipulation (Kevles, Werskey). For these reformers, eugenics spoke to women's dreams of escaping the restrictive, essentialized reproductive body, and workers' dreams of upending the hereditary distribution of power (Brittain, Haldane).

The rationalization of labor carried on in the early twentieth century in Europe, England, and America aimed at maximum efficiency by fragmenting the work process. These new industrial methods lent the goal of gaining eugenic control of the reproductive body a specifically industrial form. They treated the worker's body as a machine, breaking down the labor process into its smallest possible units, using the assembly line to enforce a uniform, external schedule, and carrying on constant surveillance. Modern literature drew metaphorically on this monitored, mechanistic, regulated, and fragmented way of life, using it to figure not just production, but reproduction. Drawing their central metaphor from Taylorism and Fordism, such representations presented a reconfigured human body—male *and* female—available for industrial production.

Just as the self-creating embryo or fetus was central to the Romantic figuration of reproduction (expressing both the power and the critique of the new civil state), so the ectogenetic fetus is central to those modern conceptions of reproduction. Ectogenesis, or gestation outside the body of a woman in an artificial uterus, figures in the most familiar literary image of reproductive technology after *Frankenstein*, Aldous Huxley's Central London Hatchery and Conditioning Centre of *Brave New World*. There, an ectogenetic assembly line of test-tube babies embodies the "principle of mass production at last applied to biology" (A. Huxley 4). Following the divine principle of "Our Ford," each bottled embryo is separated from its maternal context, placed on a conveyor belt, stimulated both chemically and environmentally, monitored and held to a normative timeframe, until the fetuses are not born but "decanted . . . as socialized human beings" (A. Huxley 8).

Six years before Aldous Huxley published *Brave New World*, his older brother Julian took a break from his zoological studies to write "The Tissue-Culture King." This science fiction short story, originally published in *The Yale Review* with the subtitle "A Parable of Modern Science," tells the story of a British medical researcher taken captive by an African tribe. Dr. Hascombe introduces the tribe to Western medical techniques—microscopy, tissue culture—and comes up with a scheme for expanding the power of the king—and coincidentally, his own power. Hascombe persuades the king to donate a sample of his own tissues, which is cultured in the laboratory, and reproduced, enabling the new doctor-priests to give each tribal member a bit of the king's immortal body, to revere endlessly. Hascombe also creates a laboratory to produce freaks and cross-breeds (chimeras), for the tribal citizens to worship. In its play on the techniques that would become foundational to contemporary reproductive technology, such as microscopy, tissue-culture, and genetic engineering, Julian Huxley's story embodies another facet of the conflicting significations attached to reproductive technology in the early years of the twentieth century.

Still another facet appeared in the debate about the social significance of extrauterine gestation catalyzed by J. B. S. Haldane's popular *Daedalus, or Science and the Future* (1923), with its speculative scenario for "ectogenesis." In one of the most striking exchanges in response to Haldane's book, a masculinist and a passionate feminist surprisingly agreed that ectogenesis could be put to feminist uses, only to disagree on the implications of that fact. Nietzschean philosopher Anthony Ludovici gloomily predicted that when extrauterine gestation became a reality, "triumphant Feminism will probably reach its zenith. . . . Men will then be frankly regarded as quite superfluous" (Ludovici 93). In contrast, novelist Vera Brittain saw ectogenesis as a temporary stage in the process of shaping pregnancy along feminist lines. She predicted that while ectogenesis would be welcomed in cases where "normal pregnancy was exceptionally inconvenient to the wife, or would involve a long separation from her husband," parents would not switch wholly to ectogenetic gestation because children thus produced would be found not to thrive. Rather, parents would find means to make "childbirth painless and pregnancy definitely pleasurable," leading "nearly all twenty-first-century parents to return to natural methods of reproduction" (Brittain 77–78). Like all images that have cultural prominence, reproductive images serve not so much to articulate a single ideological position, as to provide a site on which positions can be contested; the contrast between Brittain's and Ludovici's positions on ectogenesis reveals the power of reproductive images to express not only power, but also resistance (Poovey).

The dominant feature of postmodernism is its challenge to the master narratives of Western metaphysics and philosophy, with their bases in binary oppositions: mind/body; male/female; self/other; first world/third world; human/non-human. Depending on the definition of postmodernism to which we subscribe, the notion of an exhaustive, and therefore "true," narrative is replaced either by an emancipation from narratives that claim to possess universal truth or by a turn to the manipulation and control of information in order to maximize its efficient transfer. N. Katherine Hayles has argued that while one definition stresses philosophical emancipation and the other technological control, both descriptions of postmodernism have at their center the process of "denaturing," or "depriv[ing] something of its natural qualities."[2] Yet that very opposition between natural and denatured is itself implicated in a modernist epistemology now undermined by what Donna Haraway has called "the informatics of domination." As Haraway reminds us, "We cannot go back ideologically or materially. . . . Ideologies of sexual reproduction can no longer reasonably call on notions of sex and sex role as organic aspects in natural objects like organisms and families" (Haraway, *Simians* 162).

Science and technology have so rearranged the boundary conditions for the reproduction of human identity that the choice is no longer between the natural body and the culturally constructed body, but between different fields of bodily (re)construction bearing different social and cultural implications. The significance of postmodern representations of reproductive technology will differ, depending on the strategy deployed for denaturing the human being: whether they call into question the totalized notion of a human being (body-and-soul, the narrative of a life) in order to affirm other multiple identities and positionalities, or substitute for that totalization an instrumental focus on body fragments as segments of information subject to manipulation.

Three postmodern literary texts suggest the implications of the choice of a particular denaturing strategy. Robin Cook's *Mutation* (1989), Elizabeth Jolley's *The Sugar Mother* (1988), and Angela Carter's *The Passion of New Eve* (1977) figure images of the ectogenetic fetus, the surrogate mother, and the pregnant man. I will move from apparently complicit to resistant representations of reproductive technology—though in each novel there are gaps that figure resistance beneath complicity, or accommodation behind critique—so I will discuss these novels in reverse chronological order. But before I do so, I want to map out two analytic axes, partially overlapping, that can illuminate the denaturing process central to their postmodern representations of reproductive technology. Although these axes occupy different positions, they are also interrelated, for they measure the extent of *resistance or capitulation to* the operations of power. On the first axis, *theoretical* postmodernism aims at emancipation from all master narratives, through a constant process of undermining totalizing statements, while its opposite pole, *technological* postmodernism, works toward ever greater control of information through a process of continual mapping of chaotic systems. *Utopian* postmodernism, on the second axis, critiques the commodification that is affirmed by its opposite pole, *co-opted* postmodernism. To summarize, then, the first axis measures the extent to which knowledge production aims at emancipation or control, while the second axis gauges the extent to which material and political practices are critical of, or complicit with, existing institutionalized power relations.[3]

Robin Cook's *Mutation* stands at one end of the continuum between cultural complicity and cultural critique. Dedicated to Mary Shelley, and taking for its epigraph her question, "How dare you sport thus with life?" Cook's mass market thriller *Mutation* (1989) tells the story of Dr. Victor Frank, a researcher turned endocrinologist, who uses *in vitro* fertilization, cloning, gene therapy, and surrogacy to produce a genetically engineered genius baby—his own son, VJ. Then, in a recapitulation and conservative recuperation of Shelley's radical narrative, the monstrous son becomes a monstrous scientist, who creates a race of genetically-engineered ectogenetic fetuses in his secret basement laboratory.

In a central scene, the ectogenetic fetuses are discovered in the monster-scientist's laboratory by—and this is a crucial twist—the monster's *mother*:

> On a long bench built of rough-hewn lumber sat four fifty-gallon glass tanks. . . . Inside each one and enveloped in transparent membranes were four fetuses, each perhaps eight months old, who were swimming about in their artificial wombs. . . . They gestured, smiled, and even yawned. . . . Marsha timidly approached one of the tanks and peered in at a boy-child from closer range. The child looked back at her as if he wanted her; he put a tiny palm up against the glass. Marsha reached out with her own and laid her hand over the child's with just the thickness of the glass separating them. But then she drew her hand back, revolted. "Their heads!" she cried. . . . Marsha stared at the tiny boy-child with his prominent brow and flattened head. It was as if human evolution had stepped back five hundred thousand years. How could VJ deliberately make his own brothers and sisters—such as they were—retarded? His Machiavellian rationale made her shudder. (Cook 318, 320)

Dr. Frank has deliberately bioengineered retarded fetuses in order to breed a race of workers to help him in his grand plan: the laboratory creation, through *in vitro* fertilization, of a super-race of intellectually superior human beings. This image of ectogenetic fetuses purports to represent a technological breakthrough that threatens Western civilization. Yet the real subject is the production of political, rather than technological power: the scene yokes the unscrupulous methods of Machiavelli to the purportedly more legitimate methods of Rousseauean contractual democracy, and reinscribes the forces that consolidated the liberal state through the metaphoric construction of the autonomous civil subject. In Marsha's encounter with the ectogenetic fetuses we see dramatized: the separation of fetus and mother (as Marsha and the fetuses are kept apart by the see-through walls of the machine womb); the construction of the fetus as a miniature individual (possessed of agency and intentionality; able to watch, gesture, smile, even yawn); the construction of the mother as marginalized machine (split into Marsha-as-observer, and the gestational tanks); the appropriation of evolutionary thinking to the ends of social reconstruction (embodied in H. G. Wells's late-nineteenth-century *The Island of Dr. Moreau* and recapitulated here in the Neanderthal brows of the deliberately retarded fetuses); and finally the distinction between family ties and contractual or public relations. This last is figured when Marsha, explicitly framed as a caring woman in contrast to her son, an instrumental, rational man, wonders: "How could VJ deliberately retard his own brothers and sisters?"

As Londa Schiebinger has shown in her study of gender and early modern science, scientific truth is itself variable:

> asymmetries in social power have given great authority to the voice of science . . . science cannot be considered neutral so long as systematic exclusions from its enterprise generate systematic neglect (or marginalization) of certain subject matters and problematics. (266)

Yet although Cook's thriller dramatizes a scientific decontextualization in its plot of ectogenetic fetuses, it does so not to challenge existing categories of what is "natural," but to reinscribe them. Accepting the notion of a neutral science capable of representing reality objectively, Cook's vision of reproductive technology colludes with the asymmetrical power relations structuring that science.

While Robin Cook accepts the notion of a natural world that science can document objectively, Elizabeth Jolley uses the theme of surrogacy in *The Sugar Mother* to attack the notion that there is a natural world existing before or beyond representation. This perversely comic tale of how a young girl uses the ruse of becoming a surrogate mother to gain housing, succor, and financial assistance from a historian left alone by his obstetrician wife who is on overseas study leave, immerses us in a post-Heisenbergian world in which scientific "facts" are shaped by the very epistemological and representational processes by which they are discovered and reported.

Jolley's novel dramatizes Latour and Woolgar's influential analysis of scientific knowledge production as "a system of literary inscription, an outcome of which is the occasional conviction of others that something is a fact" (Latour and Woolgar 105). Arguing that "Scientific activity is not 'about nature,' it is a fierce fight to *construct* reality," Latour and Woolgar conclude that "science is not objective, it is projective." Instances of reality under construction proliferate in Jolley's novel. The titular theme of surrogacy is the dominant example of such a representational construction of reality, but another is the body diary kept by Edwin, readable (at different moments) as postmodern performance/body-art-as-text or a pathological textualization of a full-blown case of hypochondria:

> The books were the external, the internal and the intangible. The book of the skin (the external) had separate pages for different places on the body. He planned at some stage to have a series of maps like ordnance survey maps (in sections) of the human body, his body, with special methods of marking wrinkles, hair, moles, bruises, dry patches and the rather more unusual blemishes. Every page had its own legend and scale and he hoped, ultimately, to make an accurate index. . . . He often imagined Cecilia's pleasure at receiving the copies, handsomely bound, at some time in the future after he was dead. (Jolley 8)

Edwin's books of the body are more than merely obsessive intellectualization and sexual displacement, though they *are* that. Edwin's use of inscription to gain control over the body links the novel's themes of literature, religion, and surrogacy.

As Jolley portrays it, surrogate mothering joins two epistemological systems—the religious and the scientific—only to collapse them into a third, the discursive. Jolley's ironic representation of contemporary surrogate mothering links it to the medieval belief in the Virgin Birth; in each case a "fact" (whether scientific or religious) is revealed to be not naturally given, but socially constructed. Jolley suggests that we invented the myth of the Virgin Mary, and the new scientifically mediated role of surrogate mother, in order to cope with a set of indeterminacies and ungovernabilities: the indeterminacy at the heart of fatherhood (that paternity is always invisible, maternity visible), and the ungovernable nature of female desire and procreative power. Although both religion and science claim the authority to ascertain natural reality before or behind socially constructed appearances, in actual practice they manifest the shaping power of representation.

Denaturing is fundamental to Jolley's strategies for character creation in *The Sugar Mother.* The central couple in the novel are professionally involved in the process of fact creation and deployment: Edwin as an historian and Cecilia as a obstetrician/gynecologist and infertility specialist. Yet Jolley's narrative continually undermines the facts they manipulate professionally, as well as the facts about themselves and others that they rely on daily. Physical appearance, gender identification, sexual orientation, even their reproductive capacities and roles, are all undercut. Not only does Edwin obsessively chart in his three books of the body every physical change, inside and out, but both Edwin and Cecilia wear wigs, "for those occasions demanding change," and attend "parties . . . [that] consisted often of people who were not being themselves" (Jolley 42). Sexual identity poses problems for both characters: ambivalent and fumbling, Edwin generalizes from his own experience when he wonders about Cecilia's work on infertility, "how many of them would be fertile if they did the thing properly" (Jolley 14). And Cecilia's status as the faithful wife is complicated by her lesbian relationship with a colleague, Vorwickl, with whom she is traveling while on study leave. Even the dog, Prince, is revealed to be not male, but both female and pregnant.

All of these slyly de-authorized facts point to the central "fact" that the novel calls into question: that the young girl Leila is a "sugar," or surrogate, mother for Edwin and Cecilia's child. The doubt about the child's parentage, and the desire for paternal certainty, resonate throughout the novel: in allusions to the Virgin Birth, to the Joseph and Mary maternity wing of the hospital where Cecilia works, and to the "pensive and gentle faces of the Madonna" by Hans Memling, Durer, and Van Eyck which, as Edwin muses, "never ceased to fill him with indescribable longings" (Jolley 26–27).

When *The Sugar Mother* ends, Leila has absconded, not merely with the surrogacy payment, but with the baby too—the son she had earlier turned over to Edwin as specified in her surrogacy contract. Edwin has come to question both his paternity and Leila's "Madonna-like quality": "Like Joseph, perhaps he was not the father of this child. . . . He thought he would write something, a parable. A suburban parable for an entirely new bible" (Jolley 193). Writing her own ironic "suburban parable," Jolley emphasizes the male desire for control that connects the biblical notion of immaculate conception to the contemporary notion of surrogate mothering. She thus calls into question not just the miraculous singularity of the Virgin Birth, but also the medical-scientific distinction between a "natural" and a surrogate mother. Yet Jolley does not reinscribe the former by critiquing the latter. Rather, she reveals that all such oppositions are the product of social negotiations, for intentionality is as central to surrogacy as in ancient times the idea of fathering was to paternity (Laqueur, Strathern).

Jolley's *The Sugar Mother* illuminates the masculine anxiety produced by the invisibility of paternity. Angela Carter's *The Passion of New Eve* responds to that same male anxiety by imagining a reconstruction of the male that gives him access to woman's biological and sociocultural position and powers. Yet the resulting vision may be more disturbing than Jolley's ineffectual Edwin or Cook's scheming scientist. Jolley's drama unrolls on the purely personal level (shifting only by analogy to a mythic or religious register), while Carter explicitly uses the image of male transsexual pregnancy to connect the private realm of personal sexual relations to the public realm of the liberal civil state. With its epigraph gesturing to the social contract ("In the beginning all the world was *America*"—John Locke), Carter's novel traces a journey to the center of the sexual contract, exploring how private sexual difference is mapped onto an America whose public space is undergoing an apocalyptic unraveling.

The protagonist, a British man named Evelyn, is kidnaped and brought to the desert center of America by Mama, "the Great Parricide, . . . Castratrix of the Phallocentric Universe," in order to be made into a surgically created woman, "the new Eve." As the plan is explained to him:

> Myth is more instructive than history, Evelyn; Mother proposes to reactivate the parthenogenesis archetype, using a new formula. She's going to castrate you, Evelyn, and then excavate what we call the "fructifying female space" inside you and make you a perfect specimen of womanhood. Then, as soon as you're ready, she's going to impregnate you with your own sperm, which I collected from you after you copulated with her and took away to store in the deep freeze. (Carter 68)

The novel's emancipatory narrative challenges the identity stories fundamental to the construction of sex and gender in contemporary America, and, because of America's cultural dominance, to much of the rest of the world. Carter's tale of reproductive (re)construction is paralleled by a tale of sexual (re)construction, for Evelyn is on a quest for the woman of his dreams—the film star, Tristessa. This quest culminates when the surgically created new Eve discovers her new Adam: Tristessa, stripped naked, and unambiguously male. The secret sexual reversal that has been Tristessa's lifelong accomplishment illuminates how desire is implicated in sexual identity, making it not natural, but constructed:

> If a woman is indeed beautiful only in so far as she incarnates most completely the secret aspirations of man, no wonder Tristessa had been able to become the most beautiful woman in the world, an unbegotten woman who made no concessions to humanity. (Carter 128–29)

Shelley's unbegotten man—Frankenstein's monster—is joined now by Carter's unbegotten woman, as Carter's novel rounds back on the liberal civil state whose beginnings Mary Shelley plumbed in her nightmare vision, to give us a nightmare image of its apocalyptic demise.

The Passion of New Eve not only recalls *Frankenstein*, but anticipates Robin Cook's pulp rewrite of it, in linking the themes of single creation and surgical "birth" to cryogenics, *in vitro* fertilization, and artificial insemination. However, Carter's novel insistently deconstructs all the binary distinctions on which Cook's plot is based: male/female; public/private; science/magic; natural/cultural. Rather than preaching greater scientific control over conceptive technologies, Carter urges us to interrogate "the beginning," whether we construct it biologically (as the moment of fertilization, the origin of sexual difference) or politically (as the moment of the social contract). However we define that point of origin, Carter's novel suggests, it is not naturally given, but only *naturalized.*

Octavia Butler's *Xenogenesis* trilogy re-imagines posthuman, polysexual, interspecies reproduction—*xenogenesis*—in a plot based not on scientific procedure (like Cook) or the exploration of identity construction in society (like Jolley and Carter), but on a tried-and-true science fiction formula: the alien.[4] Kinship and difference; the alien and the familiar: these oppositions concern not just science fiction, but anthropology as well. In her study of the anthropological implications of the new reproductive technologies, Marilyn Strathern has explored their role in shaping our different conceptions of identity and relationship, inaugurating a notion of reproduction not as a given natural process, but as the outcome of an identity-constructing and/or confirming consumer choice from among a variety of identity, parental, and kinship configurations. If the historical understanding of kinship as the "social construction of natural facts" has been destabilized by the new notion of reproduction as a *choice* between various different, commodified reproductive technologies, it does not necessarily follow that identity is no longer biologically grounded (Strathern 17). Choice can mean the ability to create an elective, biologically based identity in an atomized consumer society, or it can be the decision to affirm culturally constructed identity and kinship as part of social relations.

This distinction becomes clearer when we consider Butler's oxymoronic alien, Akin, in *Adulthood Rites.* As the name slyly suggests, Akin embodies a new sort of biologically produced and culturally re-imagined kinship: genetically engineered offspring of five parents: two humans (the black woman Lilith Iyapo and an Asian man), two male Oankali, and one ooloi (the sexually neuter form of the Oankali). Amalgam of alien/kin, Akin's origins resemble those open to consumers of the new reproductive technologies, now that

> reproduction may . . . be divided among five different 'parents': two genetic parents who contribute sperm and ova for *in vitro* fertilization; the birth mother who accepts the transferred embryo, gestates the fetus, and gives birth; and the social parents who rear the child. (Henifin 1)

Akin's biologically based identity, constructed rather than naturally given, parodically represents and critiques that liberal notion of autonomous-identity-as-consumption. Yet Akin also articulates the possibility of a more politically engaged, affiliative, non-naturalized, constructed identity. Despite his greater genetic ties to the dominant Oankali, he chooses to honor first his social ties, to identify politically and emotionally with the marginal "resister humans"—that is, those human beings who resist the xenogenic plans of the new society. He thereby consciously chooses to construct his identity in sociopolitical rather than biological terms.

What is obscured by these images of the ectogenetic fetus, the surrogate mother, and the pregnant man? By their prominence in contemporary literary texts, as in the representational tradition extending back to Mary Shelley's *Frankenstein,* these three images marginalize, overshadow, or repress the pregnant female body, in all its messy, boundary-defying subjectivity. We can see the implications of the marginalization or repression of the pregnant female body if we return to Carter's *The Passion of New Eve,* which figures the most dramatically denatured of the three contemporary reproductive images: the pregnant man. Although its plot might be said to concern a man who gets pregnant, the novel ultimately represents pregnancy as occurring in, and linked to, the female body. Unlike Jolley's suburban parable with its defiant subversion of the icons of Madonna and Child, Carter's rewriting of the original narrative of Western culture (the myth of the Garden of Eden) preserves intact the central image: the female body of Eve, our first mother, thus maintaining the connection between pregnancy and the female body.

Carter does not represent that link as unquestioned and natural, however; she reinscribes it. Her protagonist, Evelyn, only becomes pregnant after he/she has been *surgically reconstructed as a woman.* Carter uses the creation of a "new Eve" to gloss the creation of the old Eve. Given Carter's deconstruction of all seemingly natural categories, we might well ask why she maintains the category of pregnant woman. This reinscription of the connection between the female body and pregnancy is integral to Carter's critique of the narrative of Western political/cultural/sociosexual arrangements. As Judith Butler has observed, "the subject . . . is constituted by the law as the fictive foundation of its claim to legitimacy" (Butler 2–3). Carter uses the representation of reproductive technology to challenge a parallel process of subject-constitution-as-institutional-legitimation. By retelling the story of Eve, she illuminates how the female subject has been constructed to authorize and legitimize the bourgeois civil state, with woman's capacity (or vulnerability) to pregnancy serving as the foundation for its sociopolitical structures. *The Passion of New Eve* reminds us that the female subject has, at least since the late eighteenth century, been constructed in and through reproduction: as politically legitimating site of deep subjectivity, as object of medical scientific knowledge and power, as machine, as monitored industrial producer, as co-opted passive consumer. ✱

Reproduction is reframed as a constructed and identity-confirming set of choices not only in Carter's *New Eve* (and Butler's *Adulthood Rites*) but also in the section on "Transsexualism and Abdominal Pregnancy" prepared as part of the discussion paper, *Developments in the Health Field with Bioethical Implications*, by the Australian National Bioethics Consultative Committee (1989). William A. W. Walters implicitly accepts the notion that identity can be (technologically) constructed when he argues that "many reassigned transsexuals would feel it was an integral part of their femininity that they should bear a child and for such people abdominal pregnancy offers the possibility that their wishes could be fulfilled." Linking "natural" and constructed sexualities, Walters's argument represents as a continuum the choices constituting one's reproductive identity, as well as specific medical techniques making such choices possible:

> [If] an abdominal pregnancy were to be successful in a biological female it might also be successful in a biological male or in a biological

> male who has been reassigned as a female. . . . It is envisaged that the embryo could be placed in a pocket of peritoneum in the omentum where it could be retained in position by suturing a flap of peritoneum over it (Walters C-11, C-20)[5]

Walters goes on to speculate that male abdominal pregnancy could be justified on the grounds of a right to reproductive autonomy: that like the heterosexual female and male, the transsexual male should be able to make an autonomous choice to engage in identity-affirming reproductive behavior.

Yet to Walters's assertion that reproductive choice functions to affirm an individual identity, Rebecca Albury counters that reproduction is a matter for the collectivity. Her different perspective suggests how the posthuman can serve to endorse not autonomous individuality but responsible collectivity. Albury acknowledges the constructed character of the transsexual body without accepting its centrality to the (re)construction of the individual identity:

> the body of the transsexual is itself a technological artifact, which has been totally transformed by a variety of social and technical practices ranging from major surgery, through the use of hormones, to makeup and dress. Serious questions need to be raised about the acceptability of including pregnancy as an inscriber of feminine identity on the formerly male body. (Albury 7)

Albury critiques Walters's uninterrogated linkage between the female body and "feminine identity," arguing instead that since "gender identity is a social practice," identity is not a matter of individual bodies or individual intentions, but rather of social bodies, social intentions, social constructions (Albury C-4). Since "individuals inscribe a (social) gender on their (natural) bodies, the question of reconstructing identity should also be decided in the social, rather than personal, arena:

> It may be that the only possible approach to the potential demand for abdominal pregnancy is to focus, not on the desires of individuals, but on the social and political questions posed by resource allocation and the ethical questions posed by the appeal to a technological imperative. (Albury C-4, C-8)

Posthumanity is not only oppressive (though it can be that), but can also affirm linkages: to other psyches, other species (animal and vegetable), and other agencies, from the technological to the multiple and intrapsychic. Physiologist J. B. S. Haldane, who first conceptualized the ectogenetic uterus in *Daedalus, or Science and the Future*, wrote a science fiction novel late in his life that mapped this affirmative form of the posthuman. The protagonist of Haldane's *The Man with Two Memories* transcends individual identity, containing within him the life experiences and memories of several people. Donna Haraway has mapped another aspect of the posthuman, in her observation, "I have always preferred the prospect of pregnancy with the embryo of another species" (Haraway, *Primate* 377). Contemporary reproductive writing, too, can articulate not only the oppressive posthuman, but also—in its resistant discourses—the new images and contexts that will shape these more productive and pleasurable models for reproducing the posthuman body.

Notes

My thanks to Ira Livingston, E. Ann Kaplan and Helen Cooper for comments on an earlier version of this essay, and to participants in "The Postmodern Body: Health, Nursing and Narrative" conference, LaTrobe University, Melbourne, Australia, June 1992, for providing the occasion for its composition.

1. The subject of Figure 6.2 is Stanford University professor Dr. Carl Djerassi, one of the inventors of the birth control pill and founder of Syntex Laboratories.
2. As N. Katherine Hayles analyzes it, denaturing has operated in the postmodern period first on language, to reveal that "signification is a construction rather than a natural effect of speaking and writing"; then on context, with the result that "contexts are increasingly seen as constructions rather than as givens"; and finally on time, which is perceived no longer as a continuum, but rather as a set of discrete and interchangeable present moments. The ultimate conclusion of this denaturing process will be the denaturing of the human, Hayles predicts, with either liberating or repressive results. Developments in both genetic engineering and reproductive technology suggest that such a denatured human being is not far in the future (Hayles 266–82).
3. "In its *theoretical* guises, cultural postmodernism champions the disruption of globalized forms and rationalized structures. In its *technological* guises, it continues to erect networks of increasing scope and power" (Hayles 291). Both terms on the second axis enact "a movement of culture and texts beyond oppressive binary categories." But utopian postmodernism critiques, while co-opted postmodernism is complicit with, "the new stage of multinational, multi-conglomerate consumer capitalism, and . . . the new technologies this stage has spawned [which] have created a new, unidimensional universe" (Kaplan 3–4).
4. My thanks to Ira Livingston for this observation (personal communication).

For discussions of Butler's earlier novels, see Donna Haraway, *Simians, Cyborgs, and Women: The Reinvention of Nature.*

5. Two works of fiction containing portraits of male pregnancy are Ryman and Gray. In the first novel, male pregnancy is plotted in terms of interspecies reproduction and communication; in the second, as part of a homosexual romance plot.

Works Cited

Albury, Rebecca. "Introduction." "Transsexualism and Abdominal Pregnancy." By William A. W. Walters. *Developments in the Health Field with Bioethical Implications.* Vol. II (April 1990). [Australian] National Bioethics Consultative Committee. C-3–C-10.

Appignanesi, Lisa. *Postmodernism: ICA Documents.* London: Free Association Books, 1989.

Belsey, Catherine. "Constructing the Subject: Deconstructing the Text." *Feminisms.* Ed. Robyn R. Warhol and Diane Price Herndl. New Brunswick: Rutgers University Press, 1991. 593–609.

Brittain, Vera. *Halcyon, or the Future of Monogamy.* London: Kegan Paul, Trench, Trubner, 1929. 77–78.

Butler, Octavia. *Dawn: Xenogenesis I.* New York: Warner Books, 1987.

———. *Adulthood Rites: Xenogenesis II.* New York: Warner Books, 1988.

———. *Imago: Xenogenesis III.* London: Gollancz, 1989.

Carter, Angela. *The Passion of New Eve.* London: Virago, 1977; rpt. 1985.

Cook, Robin. *Mutation.* New York: Putnam's, 1989.

Doray, Bernard. *From Taylorism to Fordism: A Rational Madness.* Trans. David Macey. London: Free Association Books, 1988.

Foucault, Michel. *The History of Sexuality Volume I: An Introduction.* New York: Vintage Books, 1980.

Gray, Stephen. *Born of Man.* London: GMC, 1989.

Haldane, John Burdon Sanderson. *Daedalus, or Science and the Future.* London: Kegan Paul, Trench, Trubner, 1923.

———. *The Man with Two Memories.* London: Merlin Press, 1976.

Haraway, Donna. "A Cyborg Manifesto: Science, Technology, and Socialist-Feminism in the Late Twentieth Century." *Simians, Cyborgs, and Women: The Reinvention of Nature.* London: Routledge, Chapman and Hall, 1991.

———. *Primate Visions: Gender, Race and Nature in the World of Modern Science.* New York: Routledge, 1989.

———. "Situated Knowledges: The Science Question in Feminism and the Privilege of Partial Perspective." *Simians, Cyborgs, and Women: The Reinvention of Nature.* London: Routledge, Chapman and Hall, 1991. 183–201.

Hayles, N. Katherine. *Chaos Bound: Orderly Disorder in Contemporary Literature and Science.* Ithaca: Cornell University Press, 1990.

Henderson, Andrea. "Doll-Machines and Butcher-Shop Meat: Models of Childbirth in the Early Stages of Industrial Capitalism." *Genders* 12 (Winter 1991): 100–19.

Henifin, Mary Sue. "Introduction: Women's Health and the New Reproductive Technologies." *Embryos, Ethics, and Women's Rights: Exploring the New Reproductive Technologies.* Ed. Elaine Hoffman Baruch, Amadeo F. D'Adamo, Jr., and Joni Seager. New York: Haworth Press, 1987. 1–8.

Huxley, Aldous. *Brave New World.* New York: Harper & Row, 1969.

Huxley, Julian S. "The Tissue-Culture King." *Amazing Stories* vol. 2, no. 5 (August 1927): 451–59.

Jolley, Elizabeth. *The Sugar Mother.* New York: Harper & Row, 1988.

Kaplan, E. Ann. "Introduction." *Postmodernism and Its Discontents: Theories, Practices.* London: Verso, 1988.

Kevles, Daniel J. *In the Name of Eugenics: Genetics and the Uses of Human Heredity.* Berkeley: University of California Press, 1985.

Laqueur, Thomas. "The Facts of Fatherhood." *Conflicts in Feminism.* Ed. Marianne Hirsch and Evelyn Fox Keller. New York: Routledge, 1990. 205–21.

Latour, Bruno, and Steve Woolgar. *Laboratory Life: The Construction of Scientific Facts.* Princeton: Princeton University Press, 1986.

Ludovici, Anthony. *Lysistrata, or Woman's Future and Future Woman.* London: Kegan Paul, Trench, Trubner, 1924.

Lyotard, Jean-François. *The Postmodern Condition: A Report on Knowledge.* Manchester: Manchester University Press, 1984.

Martin, Emily. *The Woman in the Body: A Cultural Analysis of Reproduction.* Boston: Beacon Press, 1987.

Mellor, Anne K. "*Frankenstein*: A Feminist Critique of Science." *One Culture: Essays in Science and Literature.* Ed. George Levine. Madison: University of Wisconsin Press, 1987. 287–312.

Pateman, Carole. *The Sexual Contract.* Stanford, CA: Stanford University Press, 1988.

Poovey, Mary. *Uneven Developments: The Ideological Work of Gender in Mid-Victorian England.* Chicago: University of Chicago Press, 1988.

Ryman, Geoff. *The Child Garden or a Low Comedy.* London: Unwin Hyman, 1989.

Sawicki, Jana. *Disciplining Foucault: Feminism, Power, and the Body.* New York: Routledge, 1991.

Schiebinger, Londa. *The Mind Has No Sex? Women in the Origins of Modern Science.* Cambridge, Mass.: Harvard University Press, 1989.

Shelley, Mary Wollstonecraft. *Frankenstein, or The Modern Prometheus.* (1818) New York: Bantam Books, 1981.

Siskin, Clifford. *The Historicity of Romantic Discourse.* New York: Oxford University Press, 1988.

Stanworth, Michelle. "Birth Pangs: Conceptive Technologies and the Threat to Motherhood." *Conflicts in Feminism.* Ed. Marianne Hirsch and Evelyn Fox Keller. New York: Routledge, 1990. 288–304.

Strathern, Marilyn. *Reproducing the Future: Anthropology, Kinship, and the New Reproductive Technologies.* New York: Routledge, 1992. *Critical Inquiry* 12 (Autumn 1985); rpt.

Walters, William A. W. "Transsexualism and Abdominal Pregnancy." *Developments in the Health Field with Bioethical Implications.* Vol. II (April 1990). [Australian] National Bioethics Consultative Committee. C-11–C-76.

Wells, H. G. *The Island of Dr. Moreau.* New York: Signet Classic, 1988.

Werskey, Gary. *The Visible College: A Collective Biography of British Scientists and Socialists of the 1930s.* London: Free Association Books, 1988.

Woll, Lisa. "The Effect of Feminist Opposition to Reproductive Technology: A Case Study in Victoria, Australia." *Reproductive and Genetic Engineering.* Vol. 5, no. 1. 21–38.

Yoxen, Edward. *The Gene Business: Who Should Control Biotechnology.* London: Free Association Books, 1986.

PART III:

QUEERING

SEVEN

The Seductive Power of Science in the Making of Deviant Subjectivity

Jennifer Terry

As part of an attempt to situate the AIDS epidemic historically and culturally, much of my work is devoted to making sense of the discursive association between homosexuality and pathology over the past one hundred years or so. I am concerned with what the confluence of these two terms—homosexuality and pathology—means for the construction of queer subjectivities particularly in the United States now, at the end of a century and a millennium. My method is to examine critically medical and scientific projects designed to gain epistemological and social control over homosexuality. I do not presume that one can do this effectively by focusing on scientific practices and discourses alone. Instead, my method is to read scientific practices as both embedded in and expressive of culturally and historically specific conditions. In short, I presume science to be situated always amongst competing meanings and explanations, and never to be in a domain free from political, economic, and cultural processes. For many readers, I am sure, understanding science *as* culture and *in* culture is a first premise, or a given of any intelligent analysis of scientific knowledge. Likewise, the notion that science is both embedded in and constitutive of dynamics of social power is by now beyond question, even (or especially) from the perspectives of many scientists themselves. But demonstrating these points seems important now at a time when the magical signs of rationality, objectivity, and scientific authority continue to shape the terms by which we imagine ourselves, our bodies, and the environments we occupy.

I would like to offer some ideas for assessing the contemporary relationship of lesbians and gay men to scientific knowledge by saying a little bit about the paradoxical history of this relationship. By now, we have voluminous evidence to show that science and medicine have played a big part in the making of homophobia. And yet, paradoxically, science has on occasion had a seductive power over gay men and lesbians especially when it purports to offer us truth, authenticity, the security of identity, and even liberation. Why and how has homosexuality become the object of scientific scrutiny? What cultural anxieties generate scientific studies of homosexuality? And why would lesbians and gay men agree to be studied by scientists? I want to pose these questions through a brief historical survey of what I think represent the main kinds of scientific research on homosexuality undertaken in the United States during this century. Hopefully, this survey will shed some light on the cultural salience of recent biological research on sexual orientation to which I will turn during the second half of the paper.

The Paradox of Seduction and Repulsion

Over the past several years I have encountered a particularly difficult and troubling problem in my research on the history of the formation of lesbian and gay subjectivities. This problem in many ways precipitated my specific interest in analyzing how science thinks about homosexuality: in trying to map what produces or constructs "queer" subjectivities in the twentieth century, I have been constantly reminded that one cannot simply disentangle the discursive conflation of homosexuality and pathology. This led me to investigate specific instances where lesbians and gay men were brought under the medical and scientific gaze (Terry 1990; 1992; forthcoming). I am immensely curious about how the processes of clinical and scientific scrutiny operate in relation to the production of queer subjectivities. And, frankly, I am most intrigued in cases where lesbians and gay men actually volunteered to be studied by experts and to be examined physically in ways that we might find utterly draconian today.

In order to take up the question of why lesbians and gay men volunteer to be studied by doctors and scientists and why some are inclined to embrace science, we need first to note how this question is itself located in a particular historical moment. Many of us now are deeply skeptical about the grandiose promises of science but for most of this century in the United States, science has held the status of being the sacred avenue to truth as well as being a source of national strength. Throughout this discussion I deliberately use the term science broadly to refer to a broad range of disciplines from the natural to the physical to the social sciences, including medicine, which share certain philosophical and methodological assumptions and approaches based on the idea that science is the privileged mode for discovering Truth. What these disciplines have in common is that they generally believe (1) in the importance of testing hypotheses through careful techniques of data gathering, experimentation, and observation; (2) that truth is empirically measurable and can be reproduced in an experimental setting; (3) that the scientist can maintain impartiality and neutrality by following certain procedures; and (4) that what scientists find in their laboratories or through their questionnaires or through their clinical observations has some utility to humankind.

Practitioners of science claim to be objective, impartial, neutral, and virtuous because they have helped us to know more about the world. Science has been heralded as a primary method for solving social problems like crime, poverty, and disease. However, now in the late twentieth century, after revelations about Nazi medicine, after the invention of the nuclear bomb and germ warfare, and after the ecological horrors wrought by technological development, many of us have developed a very skeptical view of the doctrine of rationality and its practitioners.

Ironically, coexisting alongside this general cultural skepticism about science and doubt about its ability to "discover" the truth, is a persistent faith in the science, if not in scientists. In and of itself, scientific knowledge continues to be seen as virtuous. On top of this, research projects on a grand scale—like the Human Genome Project, Star Wars, the Strategic Computing Initiative, and cancer research—are justified in terms of their essential importance to the vitality and security of human beings. So at the end of the twentieth century, the utopic and dystopic images of science exist concurrently in a curious way.

It is worth noting that skepticism toward scientists and doctors among some groups of people—namely women, people of color, poor people—is the result of a long history of abuse and neglect. But where are lesbians and gay men situated in relation to this contradictory state of horror and hope about science? We have particularly complicated histories which must be brought to bear on this question. Lesbians and gay men can thank scientists and doctors for naming them as pathological beginning in the latter half of the nineteenth century. And for much of this century we have been regarded as anomalies to be explained, if not patients to be treated. As a result, many are deeply ambivalent about the theories and applications of science to questions about sexuality.

We, as deviant subjects, have had to account for ourselves as anomalies. We are compelled to ask certain *questions of the self,* beyond the generic question of "Who Am I?" In addition we ask "How did I come to be this way?" "How and why am I different?" "Is there something wrong with me?" "Is there something in my background that would explain my homosexuality?" "Is there something different about my body?" "Am I a danger to myself or others?" Deviant subjectivity is forged in the relay of these questions, where a number of intended and unintended effects are produced. What follows is a brief sketch of several different kinds of studies in which homosexual subjects have been motivated to ask particular questions of the self—that is, to account for themselves—through the authority of science.

A Brief Survey of Scientific Studies of Homosexuality

Scientific inquiry into homosexuality is particularly interesting to analyze because, as I hope to show, each episode or mode of scientific research aimed at studying queers has been shaped by and, in turn, influences the social relations and the cultural context from which it emerged. I propose that we look at various kinds of inquiry as they encode and enact particular cultural anxieties. How have these anxieties shaped the terms of deviant subjectivity? Let's begin in 1869 when German physician Karl Westphal referred to the homosexual as a sufferer of *contrary sexual feeling* (1869). (On the basis of his observations of a girl who liked to dress like a boy and who acquired sexual satisfaction with other girls, Westphal concluded that her abnormality was congenital and thus should not be prosecuted by the police.) This is the moment when the homosexual as a particular type of person was discursively spawned. And, importantly, this naming process took place in the context of clinical medicine, casting the homosexual as a fundamentally diseased and degenerate being. Through individual case histories of patients, nineteenth-century neurologists like Krafft-Ebing and Charcot described homosexuality as a manifestation of innate degeneracy and nervousness, likening it to the morally and physically dissipating practice of masturbation (Krafft-Ebing 1886; Charcot 1882).

At the heart of much of this early scientific discourse and subsequent studies of homosexuality was a fascination and obsession with the body. Bodies, through their structural characteristics, motions, habits, and behaviors were territorialized in certain ways and treated as sources of scientific evidence about perversion. And thus they became sites of phantasmatic projections both from scientists studying those bodies and from the subjects who inhabited them. Through techniques of clinical surveillance and diagnosis, homosexual bodies as they were imagined by men of science throughout this history, were in some important way already posthuman bodies (although they have been frequently misperceived as "subhuman" bodies). That is, they were objects to be measured, zones to be mapped, and texts to be read. Machines and morals functioned instrumentally and phantasmatically to "find" the sources and traces of perversion. For it was both on the surfaces of perverse bodies and in their dark interiors that homosexual desire was presumed to originate and proliferate. Fully implicated in perverse desire and behavior, homosexual bodies in this history were figured not merely in terms of benign difference, but as dyshygienic, posed in opposition to the whole and wholesome organic body of the "human" (read: white, heterosexual gentleman). The phantasmatic homosexual body, fragmented concurrently through measuring devices and homophobic precepts, became a text of telltale signs which themselves functioned as indices of moral character. For the homosexual subjects themselves, the body was both constrained by this moral quest of science to find perversion on the body, and always excessive to any such quest. Historically, deviant subjectivity is profoundly bound up with contests about the roles, functions, possibilities, and violations performed by queer bodies.

Toward the end of the nineteenth century, physicians labored to establish the visible stigmata of homosexual degeneracy. In this way, the clinical study of perverts was linked with a larger scientific interest in classifying human cultural diversity in biological terms. Bodies had become territories for siting all kinds of differences inside and outside of Europe and the United States (Fausto-Sterling forthcoming; Gilman 1985; Gould 1981; Green 1984; Marshall 1990; Proctor 1988; Russett 1989; Sekula 1986; Stocking 1976, 1987). Homosexuals were one of a number of "internal others" within the West—alongside criminals, prostitutes, and the feebleminded—whose bodies were believed to carry the germs of ruin. In nineteenth-century Europe and America, the belief that moral character and psychical features were fundamentally tied to biology came to the fore with a vengeance at a moment of heated debate over who would enjoy the privileges of legal and economic enfranchisement in a newly reconfigured public sphere. In the United States, anxieties about the abolition of slavery in 1865 and the rise of feminist agitation for the vote fueled scientific research aimed at demonstrating that social inequality was merely a matter of biology and nature. It was around this moment when the sodomite was constructed as a threat to public hygiene and the mannish woman was characterized as a threat to the private realm of the family—nothing short of a woman on strike against marriage and motherhood. Science and medicine were installed as keepers of the public trust, and a fascination with all things modern made the idea of eugenical engineering of a population's genetic stock compelling across the political spectrum.

Interestingly the early clinical study of perverts and inverts engendered the resistant tradition of Magnus Hirschfeld who at the end of the nineteenth century began to collect his own histories of individuals who volunteered to give accounts of their own homosexuality (Hirschfeld 1914, 1975; Nunberg and Federn 1962). Hirschfeld's Institute for the Study of Sexual Science in Berlin used the idea as well as the textual conventions of the case history to document that those with contrary sexual feeling were benign natural anomalies, afflicted not by biological defects but by the social hostility that surrounded them. It is interesting to note that the methods Hirschfeld used for gathering this information resembled the traditional medical case history model, including a physical examination of homosexuals. The secrets of the self that Foucault described as central to the modern clinical confession of psychoanalysis were conveyed in the hopes that they would bring homosexual subjects a greater sense of self-knowledge (1978, 1980). At the same time, homosexuality would be revealed as benign difference rather than deviance and pathology. In this early instance of what we might call (at the risk of being "presentist") gay-positive science, Hirschfeld and those who offered up their stories believed that scientific knowledge would bring social tolerance, legal protection, and even personal liberation. Science was viewed as a powerful means for gaining visibility and eradicating prejudice.

Following the nineteenth-century clinical case history model, homosexuality in the first half of the twentieth century came to be the object of an array of behavioral surveys undertaken by biologists, sociologists, and anthropologists, often working in conjunction with psychiatric physicians. This kind of research did not wholly supplant the clinical case model because, as we know, homosexuality continued to be seen as a medical malady throughout most of this century. But this behavioral research used statistical methods to quantify the incidence and nature of homosexuality as part of a larger interest in constituting norms within the general population. These studies were tied to other normalization efforts linked to military recruitment in the First and Second World Wars, to public health campaigns related to venereal disease and eugenics, and to marital adjustment surveys aimed at policing the institutions of marriage and family (Davis 1929; Dickinson and Beam 1934; Kinsey 1948, 1953; Landis et al. 1940; Henry 1934; Henry and Galbraith 1934; Henry and Gross 1941; Strecker 1946; Strecker and Appel 1945; Terman 1938). A scientific zeal to measure everything from intelligence to vocational abilities to rates of sexual pleasure characterized the social engineering of the first half of this century. Anxieties engendered by economic depressions, waves of immigration into the United States by Southern and Eastern Europeans, and internal migration of African Americans to northern cities fueled the scientific rationale to solve social problems through techniques of quantification.

One study of homosexuality from this period combined the clinical case history method with scientific methods in a very interesting fashion. It was conducted in the late 1930s in New York City under the auspices of the Committee for the Study of Sex Variants, a group made up of twenty biological and social scientists, and doctors from various specialty areas (Henry 1941). The subjects of the study included forty lesbians and forty homosexual men who volunteered to be interviewed and to be examined physically in great detail by all kinds of different doctors, including surgeons and gynecologists.

One of the most interesting facets of the study for my purposes is that it was largely made possible by the volunteer work of a lesbian named Miss Jan Gay, who was a freelance journalist and novelist and part of the Greenwich Village lesbian scene in the 1920s and 1930s. During the 1920s, Gay traveled to Europe looking for information about homosexuality. In addition to the many libraries she consulted, Jan Gay visited Magnus Hirschfeld's Institute and gathered information about how to conduct a survey about sexuality, including what kinds of questions to ask. She put her new knowledge to work, interviewing some 300 lesbians in Paris, London, Berlin, and New York. This sample survey, which was based on a questionnaire she adapted from Hirschfeld, formed the methodological basis for the study conducted by the Sex Variant Committee during the 1930s. The interview questionnaire asked a series of questions about the subject's family background and ancestors, childhood experiences, her adult experiences, sexual desires, and opinions about society's attitude toward homosexuality. In addition, the protocol for research included various physical examinations designed to determine any physical features which distinguished homosexuals from the "general population."

How could the subjects have said yes to science in this way? It is impossible to determine with certainty what specifically motivated each of the eighty subjects to be interviewed and examined by doctors. But I do know that they were not coerced, at least not in any simple way. They received no money nor did they get any special favors in return for their participation. This put them in a different position from involuntary subjects of other scientific studies—for example, mental patients, prisoners, parolees, potential army recruits, soldiers, and reform school inmates—who were forced, more or less, to comply with the demands of probing experts.

What reward was offered the sex variant subjects, if not money or special favors and privileges? Why did they talk to psychiatrists and allow doctors to probe and measure them? There are several possible explanations. Clearly, Jan Gay was crucial in getting the eighty subjects for this study because she was part of a homosexual subculture in New York at the time. Many of these subjects were acquaintances and friends of Jan Gay and, as a favor to her, agreed to take part in the study. Beyond that simple fact, the psychiatrist who authored the volume compiling the case histories, assured readers that most of the subjects agreed to participate in the study because they believed it would advance the cause of social tolerance toward lesbians and gay men. Indeed, this motivation was more than mere conscientious duty; no doubt it inspired many of the subjects to avow homosexuality as a desirable alternative to heterosexuality. And they were able to intervene in the terms of homosexual representation at one of its most powerful points of generation—medico-scientific discourse.

But these explanations alone fail to capture the complexity of how medico-scientific discourse came to be an avenue of self-enunciation for these people who in many ways had physicians and scientists to thank for their social stigmatization. Perhaps in addition to relying upon science to defend them, the subjects volunteered to be studied because they believed that they might also learn something valuable about themselves in the process. And they might have also believed that being studied by scientists was a way of becoming visible within the larger society.

Around the same time this research was being done, Alfred Kinsey and his team of researchers were busy attempting to survey sexual behavior in the general population. Kinsey's voluminous studies published in the late 1940s and early 1950s were dedicated to discovering the truth about sexual behavior through both qualitative and quantitative methods (1948, 1953). His work is important to place in the history of scientific studies of homosexuality precisely because Kinsey did not single out homosexual people for study. Instead, he attempted to give a complete picture of the range and frequency of sexual practices throughout society, arguing that previous studies of homosexuality were deeply flawed because they presumed the homosexual to be a psychically and somatically distinct type of person. Kinsey's emphasis was on behavior, no matter what the self-identification of those who practiced it. Nor did he care about the psychological motivations for why people behaved the way they did. Subsequently, he sought to remove homosexuality from the stigma of medical diagnoses, having demonstrating that it was commonly practiced throughout the population. Kinsey's research opened up a space for thinking about homosexual practice as widespread and "natural." And, along with demolishing the idea that homosexuality is practiced only by a distinct type of person, Kinsey's work powerfully disrupted a scientific tradition of looking for signs of homosexuality in certain bodies—although only momentarily.

Immediately following the publication of Kinsey's research, a number of very vicious books were written mainly for mass audiences by psychiatrists and psychoanalysts who vehemently opposed Kinsey's claim that homosexuality was natural and widespread in the population. They used case histories of homosexual mental patients to argue that lesbianism and male homosexuality were indeed morbid pathological conditions (Bergler 1958, 1959; Bergler and Kroger 1954). Symptoms included immaturity, deception, and even treason, making the homosexual just as dangerous to the nation's security as the treacherous communist. Deeply homophobic psychiatrists like Edmund Bergler were more than willing to concede that homosexuality was not a biologically based condition, but a diseased lifestyle that should be subject to psychotherapy. In his book *Homosexuality: Disease or Way of Life,* Bergler argued that homosexuals were a small, psychotic group of people, and that it was only because homosexuality had been glamorized that these maladjusted and self-indulgent people gravitated toward it (1956). Bergler, along with Frank Caprio and Isaac Bieber, produced what we might call the xenophobic Cold War texts attacking homosexuality as a psychological condition that threatened the security of the family and the nation. A host of treatments and so-called aversion therapies were devised to treat the treacherous malady (Caprio 1955; London and Caprio 1950; Bieber 1965; Socarides 1968). Even while they argued that the body and biology had nothing to do with homosexuality, these men attempted to single out the homosexual as a pathological type of person, and thus to allay social anxieties unleashed by Kinsey's findings of widespread homosexual practice. But it was against this hostile and speculative psychoanalysis that Alfred Kinsey was able to appear as a careful, unbiased, and methodologically sound scientist among those in the early homophile organizations.

In their eyes, Kinsey stood out as the hero and, of course, the ultimate truth-teller based on the belief that "statistics didn't lie," and that the documented frequency of homosexual behavior in the population meant that it was both natural and normal. In the emergent homophile discourse of the late 1940s, one can hear the appeal of social scientific techniques of statistical analysis which rendered a broader and—in the minds of many—a more accurate picture of homosexuality. This was accompanied by explicit opposition to patient-based psychoanalytic or medical studies. Homophile activists in the Mattachine Society and Daughters of Bilitis (DOB) argued that, for one thing, the famous psychiatric studies of Bergler and others drew their conclusions about homosexuality from mental patients or people who were, in some other way, maladjusted to society. Homophile activists argued that scientists should study homosexuals who were not patients and who were adjusted in every other way. Of course, this carried with it a valorization of assimilationism, and a dependency upon science to articulate political claims for mainstream tolerance of homosexuality.

The homophile embrace of social science promoted two main points. First, it sought to argue that homosexuals were not, by definition, sick. In fact, it stressed that homosexuals were average people, just like everyone else. Obviously this emphasis banked on social conformity and resulted in the homogenization of differences among homosexuals into a model of the perfect "adjusted" homosexual. Secondly, the homophile interest in scientifically generated statistical surveys was related to arguing that homosexuals represented a minority, but a substantial one, worthy of some recognition for its social and cultural contributions. Scientific surveys became a strategy for visibility. In fact, the Daughters of Bilitis stated one of its foundational principles to be the gathering and dissemination of authoritative and reliable information about lesbians. For this they sought the expertise of sympathetic psychiatrists and social scientists.

Gradually, by about 1963, it is possible to trace the emergence of a split between various chapters of the Mattachine Society and DOB over the question of how important it was to have scientific studies conducted about homosexuality. It is interesting to note that no one was particularly in favor of biological or medical research on homosexuality at this moment. This consensus fits with that of the general public at the time when the horrors of Nazi medicine made headline news, and the whole idea of biological explanations for social inequality were coming in for harsh criticism. Frank Kameny, a gay man and an astrophysicist who fought against the homosexual purges from the U.S. government during the 1950s, argued in the pages of the DOB publication, *The Ladder,* that spending so much time and energy on scientific and psychological research was a waste of time for homosexuals (1965). Kameny argued that most of existing scientific studies lacked rigor and relied on an unsupported assertion that homosexuals were sick or defective. To counter this assertion, he insisted on a militant position that refused the notion that lesbians and gay men were sick, and argued that it was time to fight for homosexual rights and for homosexuals to speak for themselves, rather than taking the meek position of hoping that doctors and scientists would find homosexuals normal enough. The only scientific studies Kameny condoned were those that helped lesbians and homosexual men to understand their own worlds and lives better, not those which were meant to persuade the general public or other experts that homosexuals should be tolerated. Florence Conrad, chair of the DOB research committee, countered Kameny's editorial, arguing that lesbians and homosexual men needed to have their experiences translated by scientists so that other scientists and the lay-public would listen with interest and be persuaded that being gay was okay. For Conrad, militant action would only allow them to ignore and marginalize homosexuals. Instead, she recommended a careful course of cooperation.

Conrad had a great deal of faith, particularly in the virtues and possibilities of social scientific investigations (1965).

Then, in 1972, Del Martin and Phyllis Lyon published *Lesbian/Woman,* which they describe as a *subjective* account of lesbianism (1972). In their introduction, they explicitly set out to recuperate the experiences of lesbians from the distortions of most medical and scientific accounts. Martin and Lyon claimed to have produced neither a true confession nor a scientific book, but one that was written from subjective experience. They affirmed the book as partisan, not only rejecting the idea of objectivity and scientific neutrality, but also making the argument that lesbians must speak in their own terms, not through those set out in the experts' frameworks. But one of the things that is most interesting to me about *Lesbian/Woman,* a foundational text of lesbian-feminism, is how much it, in many respects, resembles previous social scientific surveys and early psychiatric case histories produced as a result of voluntary lesbian participation in studies. One can identify a similarity in the discursive structure of the subjects' self-descriptions reported in *Lesbian/Woman* and those of the early psychiatric interviews which were part of the Sex Variant study from the 1930s. Again we find the articulation of questions of the self— "What am I?" "How did I come to be this way?" "How and why am I different?" "Is there something wrong with me?" Hence lesbian-feminist discourse took on some of the same questions raised earlier by medical and scientific discourses that conflated homosexuality with pathology. This time, however, these questions provided the means for explicitly generating a counterdiscourse which replaced scientific authority with a new, authentic thing called "personal experience" in order to claim that homosexuality was healthy. Any pathology surrounding it was caused by social prejudice and homophobic and sexist hostility.

Whither the Homosexual Body?

What was happening to the scientific search for homosexuality in the body during this time? Earlier studies from the 1930s aimed at determining distinct somatic features of homosexuals for the most part failed to produce any such evidence. Most of them focused on the overall physical structure of bodies, measuring skeletal features, pelvic angles and things like muscle density and hair distribution. They hypothesized that homosexuals would show physical characteristics of the opposite sex, but none could provide conclusive evidence of this. Furthermore, endocrinological studies from the first part of this century relied on crude methods and were entirely unsuccessful in their attempts to link homosexuality with specific hormonal activity. But these failures by no means put to rest the hope of finding innate biological markers of homosexuality. Beginning in the early 1960s, a growing number of hormonal experiments were conducted on both rodents and non-human primates as part of a renewed attempt to determine the effects of hormones on sexual behavior, and especially to find a link between hormonal activity and homosexuality. Generally it was (and is still) believed that the rodent sexual behavior is more thoroughly dictated by hormonal factors so rats and mice were used to isolate the effects of castration and hormonal injections, and to develop theories about the origins of sexual orientation in the womb and in early infancy. Extrapolating from rodent experiments, scientist Gunter Dorner drew on the research of former Nazi scientist, Konrad Lorenz, to hypothesize that homosexual men had higher levels of estrogen and lower levels of testosterone than did heterosexual men because they had noticed that in mice, males who were castrated at an early age tended to exhibit "female-typical" behavior (i.e., they presented themselves to be mounted) (Dorner et al. 1975, 1983; Lorenz 1950, 1966). Similarly, homosexual behavior in female mice was believed to be the result of higher levels of androgen in the mother's womb. Implicitly defining human homosexuality as a matter of gender inversion, Dorner assumed that the human analog to the castrated, effeminized male mouse was the gay man. The idea that gay men commonly *mount* other men—mounting being the signifier of male-typical behavior in rats—was never reckoned with in these studies. To appropriate the parlance of sadomasochism, for the male rat, homosexuality meant being a bottom; for the female rat, it meant playing top.

Scientists conducting these hormonal experiments conjectured that homosexuality was a congenital condition, originating in the earliest stages of development, and arresting the natural maturation process of the male so that it behaved more like the species type; that is, more like a female. In contrast, female offspring showered by storms of androgens in the womb would falsely mature beyond their primary status to develop masculine characteristics (Downey et al. 1987; Ellis and Ames 1987; Griffiths et al. 1974; Meyer-Bahlburg 1979, 1984; Money 1987). Some psychiatrists took a keen interest in this type of research as a way of accounting for sissy-boys and tomboys (Friedman et al. 1977; Friedman and Stern 1980; Green 1987; Stoller 1987). Many acknowledged that hormonal activity could be influenced by the environment and family relations. But, to a great degree, they biologized the process of psychosexual development to emphasize that sexual orientation is deeply embedded in the body from an early age—perhaps a result either of a genetic predisposition to hormonal anomalies or the outcome of maternal stress that subverted the processes by which male and female embryos normally develop.

Interestingly, during the late 1960s and 1970s, this kind of research was roundly criticized for lacking scientific rigor as well as for its pretensions to explain complex human relations in biologically reductive terms. During this period, feminists, Marxists, and liberals decried attempts to explain social differences and inequalities in terms of biology. Hormonal research involving human subjects was seen by many as downright ghoulish, reminiscent of Nazi medicine and fundamentally retrograde because it defined homosexuality as an abnormal condition or defect. To a great degree, among liberals and radicals, scientific projects to discover the biological basis of homosexuality were regarded with no less disdain than medical efforts to treat homosexuality as a mental disorder. Feminists and gay liberationists were among the most vocal critics of biological determinism.

At this point, in the 1990s, I am interested in asking how and why, more than twenty years later, scientific research focused on the body is now being welcomed among some gay men and lesbians as a means for understanding ourselves better and for defending ourselves against growing homophobic hostility. It isn't that the scientific interest in finding homosexuality in the body ever really went away; but why is new research on the biology of homosexuality being embraced by certain gay organizations and individuals? What has made this shift in thinking possible?

Where Are We Today?

I'd like to take a moment to juxtapose two bold quotes, one from 1973, the other from 1993, to illustrate a point about the significant shift in thinking concerning the place of biology in constructing lesbian and gay subjectivities and political identities. In 1973, Anne Koedt articulated a common tenet among radical lesbian feminists at that time: "Basic to the position of radical feminism is the concept that biology is not destiny, and that male and female roles are learned—indeed that they are male political constructs that ensure power and superior status for men" (Koedt 1973). Twenty years later, in March of 1993, gay journalist, Chandler Burr, in a cover-page article in *The Atlantic Monthly* asserted a very different position, and one that seems to be oblivious to either radical or feminist critiques of biological determinism:

> Homosexuality's invitation to biology has been standing for years. Homosexuals have long maintained that sexual orientation, far from being a personal choice or lifestyle (as it is often called), is something neither chosen nor changeable; heterosexuals who have made their peace with homosexuals have often done so by accepting that premise.

> The very term "sexual orientation," which in the 1980s replaced "sexual preference," asserts the deeply rooted nature of sexual desire and love. It implies biology. (Burr 1993)

Maybe Chandler Burr is correct in reporting the perspectives of *some* gay men in establishing biological explanations for homosexuality, and beseeching scientists to study us more. But Burr's assertion stages a false consensus on the matter among lesbians and gay men. Some gay men's narratives of "having always felt this way" are very powerful indeed at the subjective level. And they can also perform the rhetorical function (albeit fundamentally defensive) of telling homophobes to "fuck off." But I think it is important to note that many women, in particular, feel a great deal of ambivalence about grounding identity and personal narratives in biological difference since biological explanations have been deployed historically to keep women in a subordinate position to men. Likewise there is considerable resistance among many lesbians and gays of color to the idea that homosexuality is biologically based, again because biological explanations have been largely in the service of marginalizing certain groups defined as naturally inferior to white men (Hammonds 1993).

Perhaps it would be useful to take a look at the larger cultural and historical context out of which this new biological evidence of homosexuality is emerging in order to make sense of this shift from radical feminism in 1973 to gay rights politics in the 1990s. It strikes me that there are several key cultural and political developments that make the 1990s a very different place to be than the 1970s. To begin, let's consider some of the changes in the relation of scientific knowledge to society generally, and then try to analyze how these position lesbians, gay men, and queers today.

We are living now in the age of the magical sign of the gene. There is a great deal of hope riding on this "holy grail" of the late twentieth century. Scientists promise that if we can figure out the exact function and location of genes within the human body, the human population could be rid of diseases and defects. And even more compelling, knowledge of genetics is marketed as an avenue for self-knowledge—knowledge of our proficiencies, our possibilities, our limits, our histories and our futures. We are told by scientists working on the Human Genome Project that genes can explain to us who we are at the most fundamental level of DNA. Lobbyists for the Human Genome Project (with its present annual budget of $135 million) market this new "Manhattan Project" on the one hand, as a means to both unify humans as a population sharing many genetic traits, and on the other, as a means of making distinctions between types of people. No doubt, this latter option offers great appeal among insurance companies and employers who would like to be able to be deny coverage to those who have "pre-existing" (i.e., genetic dispositions to) diseases. Likewise, people like Frederick Goodwin, head of the National Institutes of Mental Health, are interested in locating the genetic and neurochemical bases for violence, and propose the screening of inner-city children who seemed to be "incorrigible" to see if their bodies house the evil seed. Genetic explanations for social inequalities are extremely attractive in a time when the welfare state is in decline and the brutality of poverty diminishes the life expectancy of an entire generation of children of color living in our cities.

The promises of genetics are grandiose. Not only will the world be rid of disease, but knowledge of genetics will help us to maximize biological resources at a moment of fear over global agricultural scarcity. For Americans, genetic research promises to do even more than fortify our human and natural resources: it promises to save our economy in the face of fierce global competition, a campaign headed up by Dan Quayle. Biotechnology is to the 1990s what nuclear weapons development was to the 1960s—the putative guarantor of America's economic and political influence over the destiny of the planet. Never mind that metanational corporations dealing in biotechnology and genetic research will be selling our genes back to us once they isolate and patent key fragments.

I mention the magical sign of the gene and its political economy because over the past several years two scientific teams have reported a "genetic" basis for homosexuality. In late 1991, psychologist Michael Bailey and psychiatrist Richard Pillard reported that among the identical twins they studied, about half identified themselves as either gay or bisexual. A smaller figure of about twenty-two percent of genetically related brothers they studied were both gay or bisexual, and an even smaller number of about eleven percent of those who were raised as brothers through adoption and thus who were not related genetically identified themselves as either gay or bisexual. Even with these meager findings, the headlines cried out: "Scientists Find That Homosexuality is Genetic." Bailey and Pillard's research was full of methodological problems too numerous to recount here. But the most troubling was that they gave no explanation as to how they were using or measuring the term *sexual orientation.* The categories of homosexual, heterosexual, and bisexual were taken on face value as the subjects defined themselves, as if we (or they) all agree on the meanings of these terms. Furthermore, the study was not based upon random sampling techniques but recruited its subjects through gay newspapers, thus effectively weeding out men who may engage in homosexuality from time to time but do not read gay magazines or would be loathe to answer such an advertisement. Because the research required the cooperation of the gay subjects' brothers, it also weeded those men who were not out of the closet to their families or who came from homophobic families with brothers who would never agree to be part of such a study. Using the same problematic methods, the later study of female identical twins produced virtually the same statistical findings (Bailey, Pillard et al. 1993). But even with these problems of method and conjecture, both studies were touted as evidence for a genetic basis for homosexuality because the concordance rate for sexual orientation was higher in identical twins, or those who shared the same genetic material. Each pair of identical twins was reared together and yet no method was used for determining the influence of social environment and familial relations on sexual orientation. To add to the methodological weaknesses of the study, a good half of the identical twins did *not* show concordance for sexual orientation.

It is interesting to note that neither of the researchers involved in the so-called "gay twins" studies were geneticists or molecular biologists, in spite of the media representations of them. The twin-studies researchers took neither tissue nor blood samples of subjects to analyze the DNA or genetic material. Instead, it was on the basis of subjects' self-identification that the researchers determined that homosexuality was genetic, but in no more than fifty percent of the cases. The actual properties of the participants' DNA did not even enter into the discussion. Nevertheless, the magical sign of the gene was invoked to make sense of this research, and to represent the researcher as an engaged scientist. Indeed, in the media blitz surrounding this research, psychiatrist Richard Pillard was featured in *Newsweek* holding the magical object of his study, a molecular model, mimicking the traditional iconography of great men of science from Copernicus to Watson and Crick, in spite of the fact that his study was only "genetic" in the crudest sense of the word (Gorman 1991). Perhaps it would have been more accurate for him to be holding a copy of a gay newspaper from which his subjects were recruited.

What else has gone on in science and culture in the last twenty years making this new research on sexual orientation possible? In 1973, when Anne Koedt made her impassioned statement denouncing biology as destiny, no one had ever heard of AIDS. Since then, the AIDS epidemic has profoundly and devastatingly transformed the nature of lesbian and gay life in the United States. Our relations with one another, our understandings of ourselves, our sense of sexual possibilities, and our ideas about political mobilization have undergone massive transformations in the face of this deadly virus and the social neglect and homophobic contempt that have accompanied it. Our bodies are bound up in medical discourse and practices, once again, but this time under new, urgent, and

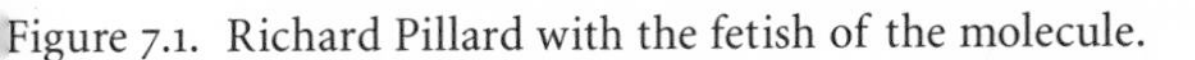
Figure 7.1. Richard Pillard with the fetish of the molecule.

deadly conditions. And these new conditions produce new ways of imagining the body in relation to subjectivity. It is not surprising that the privileged domain of the body where our innermost secrets and sexual passions are thought to reside is being imagined as a source of meanings in the face of this social atrocity. These days, even as they are theoretically and materially disintegrating, we imagine our bodies as a point of origin for exploring contemporary and very pressing questions of the self. Neuroscientist Simon LeVay's own story of what compelled him to undertake research on sexual orientation is a tale of grieving, of trying to make sense of himself as a gay man in the face of deep depression about the loss of his lover to AIDS. By his own account, LeVay's shift in focus from work on the neuroanatomy of vision to the neuroanatomy of sex and sexual orientation was a crucial part of his recovery process (LeVay 1993; Dolce 1993).

And there is a more palpable, material relationship between the epidemic and much of this new research. Indeed, AIDS provided LeVay with the very brain tissue he used to conduct his research on the hypothalamus (1991). It was men who died of AIDS who constituted the overwhelming majority of his subject population, and it was their autopsied brain tissue he used to produce his distinction between the categories of homosexual and heterosexual upon which his findings are based. What was the basis LeVay used for determining which tissue belonged to homosexual and which to heterosexual men? A single line in the subject's medical charts stating his mode of HIV transmission became the grounds for classifying a man as either gay or not. Those whose charts indicated the mode of transmission was "male-to-male" sexual contact were defined as gay, and those with other modes of transmission (IV drug use, blood transfusions, etc.) were, by default, presumed to be heterosexual. Of course, these other cases might just as well have been men who occasionally engaged in homosexual sex but who reported a different mode of transmission for whatever reasons. In other words, the journey of the human immunodeficiency virus was relied upon to account for the complexity of these men's sexual subjectivities in a masterful instance of scientific reductivism. Were it not for the early deaths of gay men through HIV infection, together with the ensuing epidemiological protocol of documenting modes of transmission in one's medical chart, LeVay's study could not have been conducted. My main point here is that AIDS provided the actual bodies for the hypothalamus study, and it provided a way to classify those bodies. It also provided the impetus for Simon LeVay to recover from his depression through the healing power of neuroscientific research, during what George Bush (the president who brought us Dan Quayle) officially proclaimed the Decade of the Brain.

AIDS also made possible the now-famous chromosomal study reported in July of 1993 (Hamer et al. 1993). Researchers at the National Cancer Institute (NCI) reported the discovery of DNA markers linking male homosexuality with a region on the X chromosome, the chromosome boys get from their mothers (prompting the facetious t-shirt I saw at the gay beach in Provincetown last summer: "Love you, Mom. Thanks for the genes."). Unlike the twins studies, this one actually did involve blood samples but, again, relied mainly on the self-reporting of gay volunteers who recounted a greater number of lesbians and gay men on their mothers' side of the family than their fathers'. Although it was no doubt colored by the fact that in general in American culture, many of us know much more about our mother's family than our father's, this self-reporting led researchers to look for the marker of homosexuality on that gift from mom, the X chromosome. This study, like the others before it, was not based on random sampling so there is no way of knowing how often this marker exists among men who practice homosexuality often or seldom, but would never identify as gay. But the relationship of this study to the AIDS epidemic is worthy of note: the money used to fund this research had been earmarked for NCI research on Kaposi's Sarcoma (KS) and lymphoma, two of the HIV-related opportunistic conditions that appear more often in men, regardless of their sexual orientation, than in women. Basically, the National Cancer Institute originally wanted to find out whether there were

Figure 7.2. Simon LeVay with the fetish of the brain.

genetic factors involved in the susceptibility to these cancers; by funding a study on the genetic markers for homosexuality, the NCI essentially presumed there was a biological relationship between KS, lymphoma and male homosexuality—rather than conditions suffered more often by HIV infected men of any sexual orientation than women. In this move, the NCI researchers reiterated—perhaps inadvertently—the idea that AIDS is a gay disease that affects particular types of people who are genetically predisposed to it. As historian of medicine Evelynn Hammonds has noted, researchers at the NCI and the Human Genome Project have money to study genetic predispositions; now they are looking for problems to solve (1993). KS and homosexuality are just two of those "problems."

There is yet another way that AIDS figures into this new cultural and scientific context, and it has to do with the nature of the current homophobic backlash against gays and lesbians. Right-wing Christian fundamentalists have declared that homosexuality is to the 1990s what abortion was to the 1980s: the enemy in a battle of moral cleansing to determine the future of the world. Recently I attended a lecture at Ohio State University sponsored by the Fellowship of Christian Students entitled "Gay Agony: Can Homosexuals Be Healed?" The guest speaker, fresh from an appearance on Pat Robertson's "700 Club," began his lecture by saying that the compulsive disorder of homosexuality has brought AIDS into the world. But it isn't too late to change, to recover from this compulsion, to turn to others for fellowship and guidance, and to overcome this deeply rooted sexual addiction. Throughout the entire lecture, homosexuality and AIDS were virtually synonymous.* The presence of such a lecture illustrates a crucial point: AIDS provides a rationale for both this kind of homophobia and a gay rights opposition to it. Indeed, the good Christian proclaimed that, contrary to what the liberal-dominated media says, there is no sound "glandular" [*sic*] or genetic evidence of the immutability of homosexuality. Beseechingly, he repeated, "you can change, you can change." It is in the face of this hostile homophobia, dressed up like Christian compassion, that LeVay's and Hamer's NCI research are being proposed as tools of political opposition. But the limitations of the gay rights-through-biology defense are striking: "Biology makes us act this way. We can't be cured. We can't seduce your children." Talk about surplus powerlessness (Minkowitz 1993).

In addition to a decade of governmental neglect and indifference about AIDS, we are now faced with a growing grass roots backlash of significant proportions against lesbians and gay men. The public displays of homophobia bring rewards these days: from Pat Buchanan's self-satisfied homophobia

*Lesbians were nowhere mentioned but everywhere implicated in the Christian Fundamentalist conflation of homosexuality with disease, moral degeneracy, and death.

at the 1992 Republican Convention, to the officially sanctioned brutality toward gay men and lesbians in the military, to the staggering rise of homophobic bashings in city streets, to the suspension of child custody for lesbian mothers around the country, to the local campaigns against lesbian and gay antidiscrimination laws: it is quite clear that lesbians and gay men are surrounded by growing numbers of enemies. And it is in these times of defensiveness and of feeling beleaguered that our bodies, presented to us through the authority of science, appear to be refuges for staking a desperate claim for tolerance. As if the knowledge of a gene for homosexuality would stop the basher's club from crashing down on our heads.

An idiosyncratic reading of civil rights law provides the backdrop to why hope is invested in the biological proof of homosexuality. Now we have gay scientists and some gay leaders arguing that homosexuality is an immutable characteristic, like race, and thus homosexuals ought to be protected from discrimination. What a strange and scientifically unsupported notion of race. And what a strange and limited reading of the current status of racial minorities and of the gains of the civil rights movement which, after all, focusing its efforts on grass roots actions as well as public marches and demonstrations, relied relatively little on the courts to demand an end to racism, and not at all on valorizing the biological immutability of race. The civil rights movement was most effective in the streets through valorizing cultural diversity, not by African Americans begging to be seen as biologically different. In the 1960s, biological arguments about race had long been seen as the handmaidens of racism, just as those about gender were identified to be a central part of the architecture of sexism.

The argument for homosexual immutability betrays a misreading of the scientific research itself. Nothing in any of these studies can fully support the idea that homosexuality is biologically immutable; each study leaves open the possibility that homosexuality is the result of a combination of biological and environmental factors, and several suggest that homosexuality may be tied to a predisposition in temperament that could manifest in a number of ways (LeVay 1993; LeVay and Hamer 1994). All agree that biological, social, and psychological factors interact to produce and change the signs of homosexuality. Furthermore, these studies cannot comment effectively on the frequency of homosexuality in the general population. Nor do they offer much in the way of understanding women's complex relationship to questions of sexuality in general, let alone sexual orientation, growing up in a culture where almost every sexual act is put off limits as potentially damaging to one's feminine reputation. It is small wonder why many lesbians' stories about their own homosexual desire are not lodged in a prenatal moment nor even in early childhood when boys first taste the privilege of exclusive prerogative over sexual expression. More often, lesbians describe their homosexuality as originating in a process of feminist politicization or in the context of an adult situation where they began to recognize a desire to desire other women.

Sometimes I think this new affinity and hope invested in the biology of homosexuality is the swan song of economically comfortable white men (quietly lip-synched by straight, politically liberal essentialists) who are the only ones who know the lyrics to each verse, and who, but for their homosexual desire, have many reasons to regard science and social order with great affection. Maybe biology is a more comforting way to narrate their desires than to make sense of them in terms of cultural and historical contradictions, conflicts, and contingencies.

I want to close by coming back to the question of the relationship between scientific knowledge and sexual subjectivity. I have tried to suggest that, throughout this century, the nature of this relationship has been as political as it has been personal. That is, different queer people have thought about scientific knowledge in different kinds of ways, some more likely to be swept up by its promises and reassurances than others. Their personal stakes have varied. And each of the episodes I described here, including the present, scientific inquiry about homosexuality, is politically situated in relation to cultural anxieties. One way to shore up anxieties has been to insist that there be a definitive line drawn between "normal" heterosexuals and diseased (or, if we are lucky, merely anomalous) homosexuals. This is the dynamic in relation to which deviant subjectivity has been largely—but not thoroughly—fashioned. Thus we can read the recent scientific studies, produced by professional gay activists like LeVay and Pillard, as expressions of a kind of separatism that finds power through claiming biological uniqueness (Pillard 1992; Dolce 1993). The idea of biology being destiny is less chilling, and perhaps even liberating, to these gay scientists. Indeed, taking into consideration what Donna Haraway has noted generally and what Evelyn Fox Keller has pointed out in reference to molecular biologists, I believe we would do well to recognize that scientists—gay, straight, or otherwise—have an intimate relationship to a newly imagined "nature" in the laboratory in this final decade of the late twentieth century (Haraway 1991, 1992; Keller 1992). Particularly in the genetics laboratory, where genes are engineered and bodies can be elementally reconfigured, scientists may feel that "nature" really is more liberating if only because it is more manipulable than ever before. This sense of animated and manipulated nature no doubt acquires some of its appeal due to its historical location in relation to the decline of the welfare state. Such a decline, experienced in the aftermath of anti-social cynicism characteristic of the Reagan/Bush era, serves to underscore a sense that the "nurture" side of arguments about causality and amelioration is either passé or moot. Indeed, what LeVay and Hamer have in common with their scientific critics is the assumption that "biology" and "environment" interact in a mutually contingent and transforming manner. (And it is important to note that the term "environment" is inclusive of everything from mundane behavioral stimuli to interpsychic family dynamics, to schooling patterns, to the complexities of socioeconomic class.) Thus, with the new and increasingly popular *interactionist* model, simple biological determinism has been replaced by a much more lively and slippery understanding of nature and culture as mutually determining, representing what some have called a new wave of sociobiology for the late 1990s.

Among those gay men who are economically and socially powerful in the world, conceding that nature makes them gay is apparently less damaging than it might seem to working-class gay teenagers. A gay social worker who works with suicidal teens told me recently that the biology-is-destiny line can be deadly. Thinking they are "afflicted" with homosexual desire as a kind of disease or biological defect, rather than it being a desire they somehow choose, is, for many gay teenagers, one more reason to commit suicide rather than to live in a world so hostile to their desires (Aqueno 1993).

In *The Epistemology of the Closet,* Eve Sedgwick invites us to "denaturalize the present" and to call into question any idea that "homosexuality as we know it today" is singular, knowable, or unified. Instead, she is interested in the "performative space of contradiction" in what could be our present understandings of homosexuality; she wants to bring out the multiplicity of narratives of "homosexuality as we know it today" (Sedgwick 1990). What would it mean if "homosexuality as we know it today" became reduced in the popular imagination to a strip of DNA, or to a region of the brain, or to a hormonal condition? What would we lose in the defensive move to believe science to be our rational savior and to base our politics in biology? What does science do *for* us? What does it do *to* us? And where can we turn for new questions of the self and new ways of *performing*—as opposed to biologically manifesting—deviance?

Author's note

Thanks to Jacqueline Urla, Donna Haraway, Donna Penn, Evelynn Hammonds, Carole Vance, Jeffrey Escoffier, Patricia White, Kay Diaz, Carolyn Dean, Sharon

Ullman, Leslie Camhi, Judith Halberstam, and Ira Livingston for intellectual and personal insights that helped in the formation of this analysis.

Works Cited

Aqueno, Frank. Remarks at Out/Write Conference on Lesbian and Gay Writing and Publishing. Boston, Massachusetts (October 3, 1993).

Bailey, J. Michael, and Richard C. Pillard. "A Genetic Study of Male Sexual Orientation," *Archives of General Psychiatry* 48 (1991): 1089–96.

———, Richard C. Pillard, Michael C. Neale, and Yvonne Agyei. "Heritable Factors Influence Sexual Orientation in Women." *Archives of General Psychiatry* 50 (1993): 217–23.

Bergler, Edmund. *Counterfeit Sex: Homosexuality, Impotence, Frigidity.* New York: Grune & Stratton, 1958.

———. *Homosexuality: Disease or Way of Life.* New York: Hill & Wang, 1956.

———. *One Thousand Homosexuals: Conspiracy of Silence, or Curing and Deglamorizing Homosexuals?* Paterson, N.J.: Pageant Books, 1959.

———, and W. Kroger. *Kinsey's Myth of Female Sexuality: The Medical Facts.* New York: Grune & Stratton, 1954.

Bieber, Isaac. "Clinical Aspects of Male Homosexuality." *Sexual Inversion: The Multiple Roots of Homosexuality.* Ed. Judd Marmour. New York: Basic Books, 1965.

Burr, Chandler. "Homosexuality and Biology." *The Atlantic* (March 1993): 47–65.

Caprio, Frank S. *Female Homosexuality.* London: Peter Owen, 1955.

Charcot, Jean-Martin, and Valentin Magnan. "Inversions du sens genital et autres perversions genitales." *Archives de Neurologie* 7 and 12 (January-February 1882; November 1882): 55–60; 292–322.

Conrad, Florence. "Research Is Here to Stay." *The Ladder* (July–August 1965): 15–21.

Davis, Katherine Bement. *Factors in the Sex Lives of Twenty-Two Hundred Women.* New York: Harper and Row, 1929.

Dickinson, Robert Latou, and Lura Beam. *The Single Woman: A Medical Study in Sex Education.* Baltimore: Williams & Wilkins, 1934.

Dolce, Joe. "And How Big Is Yours?" interview with Simon LeVay, *The Advocate* (June 1, 1993): 38–44.

Dorner, Gunter, et al. "A Neuroendocrine Predisposition for Homosexuality in Men." *Archives of Sexual Behavior* 4 (1975): 1–8.

———, et al. "Stressful Events in Prenatal Life of Bi- and Homosexual Men," *Experiments in Clinical Endocrinology* 81 (1983): 83–87.

Downey, Jennifer, et al. "Sex Hormones in Lesbian and Heterosexual Women." *Hormones and Behavior* 21 (1987): 347–57.

Ellis, Lee, and M. Ashley Ames. "Neurohormonal Functioning and Sexual Orientation: A Theory of Homosexuality-Heterosexuality." *Psychological Bulletin* 101.2 (1987): 233–58.

Fausto-Sterling, Anne. "Gender, Race and Nation: The Comparative Anatomy of

'Hottentot' Women in Europe, 1815–1817." *Deviant Bodies.* Ed. Jennifer Terry and Jacqueline Urla. Bloomington: Indiana University Press, 1995.

Foucault, Michel. *The History of Sexuality: An Introduction.* Trans. Robert Hurley. Vol. 1. New York: Vintage Books, 1978.

———. "The Confession of the Flesh." *Power/Knowledge: Selected Interviews and Other Writings by Michel Foucault, 1972–1977.* Ed. Colin Gordon. New York: Pantheon Books, 1980.

Friedman, Richard C., et al. "Hormones and Sexual Orientation in Men." *American Journal of Psychiatry* 134 (1977): 571–72.

———, and Leonore O. Stern. "Juvenile Aggressivity and Sissiness in Homosexual and Heterosexual Males." *Journal of the American Academy of Psychoanalysis* 8 (1980): 427–40.

Gilman, Sander L. *Difference and Pathology: Stereotypes of Sexuality, Race, and Madness.* Ithaca: Cornell University Press, 1985.

Gorman, Christine. "Are Gay Men Born That Way?" *Newsweek* (September 9, 1991): 48.

Gould, Stephen J. *The Mismeasure of Man.* New York: Norton, 1981.

Green, David. "Classified Subjects—Photography and Anthropology: The Technology of Power." *Ten/8* 14 (1984).

Green, Richard. *The 'Sissy Boy Syndrome' and the Development of Homosexuality.* New Haven: Yale University Press, 1987.

Griffiths, P. D., et al. "Homosexual Women: An Endocrine and Psychological Study." *Journal of Endocrinology* 63 (1974): 549–56.

Hamer, Dean H., Stella Hu, Nan Hu, Angela Pattatucci, and Victoria Magnuson. "Evidence for Homosexuality Gene." *Science* 261 (July 16, 1993): 291–92.

Hammonds, Evelynn M. Remarks at Out/Write Conference on Lesbian and Gay Writing and Publishing. Boston, Massachusetts (October 1993).

Haraway, Donna. "A Cyborg Manifesto: Science, Technology, and Socialist-Feminism in the Late Twentieth Century." *Simians, Cyborgs, and Women: The Reinvention of Nature.* London: Routledge, 1991.

———. "The Promise of Monsters: A Regenerative Politics for Inappropriate/d Others." *Cultural Studies.* Ed. Lawrence Grossberg, Cary Nelson, Paula Treichler. New York: Routledge, 1992.

Henry, George W. "Psychogenic and Constitutional Factors in Homosexuality: Their Relation to Personality Disorders." *Psychiatric Quarterly* 8 (1934): 243–64.

———. *Sex Variants: A Study in Homosexual Patterns.* Volumes 1 and 2. New York: Hoeber, 1941.

———, and Hugh Galbraith. "Constitutional Factors in Homosexuality." *American Journal of Psychiatry* 13 (1934): 1249.

———, and Alfred Gross. "The Homosexual Delinquent." *Mental Hygiene* 25 (1941): 420–22.

Hirschfeld, Magnus. *Die Homosexualität des Mannes und des Weibes.* Berlin: Louis Marcus, 1914.

———. *Sex in Human Relationships.* Trans. John Rodker. New York: AMS Press, 1975.

———. *Sexual Anomalies and Perversions.* New York: Random House, 1942.

Kameny, Frank E. "Does Research into Homosexuality Matter?" *The Ladder* (May 1965): 14–20.

Keller, Evelyn Fox. "Nature, Nurture, and the Human Genome Project." *The Code of Codes: Scientific and Social Issues in the Human Genome Project.* Ed. Daniel J. Kevles and Leroy Hood. Cambridge: Harvard University Press, 1992.

Kinsey, Alfred, et al. *Sexual Behavior in the Human Female.* Philadelphia: Saunders, 1953.

———. *Sexual Behavior in the Human Male.* Philadelphia: Saunders, 1948.

Koedt, Anne. "Lesbianism and Feminism." *Radical Feminism.* Ed. Anne Koedt, Ellen Levine, Anita Rapone. New York: Quadrangle Books, 1973.

Landis, Carney, Agnes T. Landis, M. Marjorie Bolles, Harriet F. Metzger, Marjorie Wallace Pitts, D. Anthony D'Esopo, Howard C. Moloy, Sophia J. Kleegman, Robert Latou Dickinson. *Sex and Development: A Study of the Growth and Development of the Emotional and Sexual Aspects of Personality Together with Physiological, Anatomical, and Medical Information on a Group of 153 Normal Women and 142 Female Psychiatric Patients.* New York: Hoeber, 1940.

LeVay, Simon. "Evidence for Anatomical Differences in the Brains of Homosexual Men," *Science* 253 (1991): 1034–37.

———. *The Sexual Brain.* Cambridge: MIT Press, 1993.

———, and Dean H. Hamer. "Evidence for a Biological Influence in Male Homosexuality," *Scientific American* (May 1994): 44–49.

London, Louis S., and Frank S. Caprio. *Sexual Deviations: A Psychodynamic Approach.* Washington, D.C.: Linacre Press, 1950.

Lorenz, Konrad. "The Comparative Method in Studying Innate Behavior Patterns." *Animal Behavior* (1950): 221–68.

———. *On Aggression.* Trans. Marjorie Kerr Wilson. New York: Harcourt, Brace & World, 1966, c. 1963.

Marshall, Stuart. "Picturing Deviancy." *Ecstatic Antibodies: Resisting the AIDS Mythology.* Ed. Tessa Boffin and Sunil Gupta. London: Rivers Oram Press, 1990.

Martin, Del, and Phyllis Lyon. *Lesbian/Woman.* San Francisco: Glide Publications, 1972.

Meyer-Bahlburg, Heino F. L. "Psychoendocrine Research on Sexual Orientation: Current Status and Future Options." *Progress in Brain Research.* Ed. G. J. DeVries et al. Amsterdam: Elsevier Science Publishers, 1984.

———. "Sex Hormones and Female Homosexuality: A Critical Examination." *Archives of Sexual Behavior* 8(2) (1979): 101–19.

Minkowitz, Donna. "Trial by Science: In the Fight Over Amendment 2, Biology Is Back and Gay Allies Are Claiming It." *Village Voice* (November 30, 1993): 27–30.

Money, John. *Gay, Straight, and In-Between: The Sexology of Erotic Orientation.* New York: Oxford University Press, 1988.

———. "Sin, Sickness, or Status?: Homosexual Gender Identity and Psychoendocrinology." *American Psychologist* 24.4 (April 1987): 384–99.

Nunberg, Herman, and Ernst Federn, eds. *Minutes of the Vienna Psychoanalytic Society, Volume 1, 1906–1908.* Trans. M. Nunberg. New York: International Universities Press, 1962.

Pillard, Richard. "Just What Do Gay Twins Reveal?" *The Guide* (Boston) 12(2) (February 1992): 24–28.

Proctor, Robert. *Racial Hygiene: Medicine under the Nazis.* Cambridge: Harvard University Press, 1988.

Russett, Cynthia Eagle. *Sexual Science: The Victorian Construction of Womanhood.* Cambridge: Harvard University Press, 1989.

Sedgwick, Eve Kosofsky. *The Epistemology of the Closet.* Berkeley: University of California Press, 1990.

Sekula, Allan. "The Body and the Archive." *October* 39 (1986): 3–64.

Socarides, Charles W. *The Overt Homosexual.* New York: Grune & Stratton, 1968.

Stocking, George W., Jr. *Race, Culture, and Evolution: Essays in the History of Anthropology.* London: Collier-Macmillan, 1976.

———. *Victorian Anthropology.* New York: Free Press, 1987.

Stoller, Robert C. "Boyhood Gender Aberrations: Treatment Issues." *Journal of the American Psychoanalytic Association* 26 (1978): 541–58.

Strecker, Edward. *Their Mother's Sons: The Psychiatrist Examines an American Problem.* New York and Philadelphia: Lippincott, 1946.

———, and Kenneth Appel. *Psychiatry in Modern Warfare.* New York: Macmillan, 1945.

Terman, Lewis M. *Psychological Factors in Marital Happiness.* New York: McGraw-Hill, 1938.

Terman, Lewis M., and Catherine Cox Miles. *Sex and Personality: A Study in Masculinity and Femininity.* New York: McGraw-Hill, 1936.

Terry, Jennifer. "Lesbians under the Medical Gaze: Scientists Search for Remarkable Differences." *Journal of Sex Research* 27.3 (1990): 317–40.

———. "Anxious Slippages between 'Us' and 'Them': A Brief History of the Scientific Search for Homosexual Bodies." *Deviant Bodies.* Ed. Jennifer Terry and Jacqueline Urla. Bloomington: Indiana University Press, 1995.

———. *Siting Homosexuality: A History of Surveillance and the Production of Deviant Subjects (1935–1950).* Ph.D. diss., University of California, Santa Cruz, 1992.

———. "Theorizing Deviant Historiography." *differences* 3.2 (1991): 55–74.

Von Krafft-Ebing, Richard. *Psychopathia Sexualis, mit Berucksightigung der contraren Sexualempfindung: Eine klinisch-foresische Studie.* Stuttgart, 1886.

Westphal, Karl Friedrich Otto. "Die kontrare Sexualempfindung: Symptom eines neuropathologischen (psychopathischen) Zustandes." *Archiv für Psychiatrie und Nervenkrankheiten* 2 (1869): 73–108.

EIGHT

Phantom and Reel Projections: Lesbians and the (Serial) Killing-Machine

Camilla Griggers

Lizzie Borden took an ax
And gave her father forty whacks,
When she saw what she had done,
She gave her mother forty-one.

Depredatory: destructive, consuming, wasteful, deleterious; that preys upon other animals: creature of prey and its organs of capture.
Depredate: to prey upon, make prey of, to consume by waste.
Depredation: the action of making prey of, plundering, ravaging; consumption or destructive waste of the substance of anything; destructive operations.
Depredator: one who, or that which, preys upon or makes depredations, a plunderer; one who lays waste.

—O.E.D.

Deterritorializations of the feminine in postmodern cultural formations cannot be adequately calculated without some attempt to map the breakdowns occurring in the contemporary collision course between the two concurrent yet distinct cultural flows of becoming woman and becoming depredatory. Here I map the dynamics of that collision in regard specifically to the process of becoming-lesbian in the public sphere and to the corresponding process of the public lesbian becoming killing-machine.

Becoming *killing-machine* suggests the ways in which posthuman machinic assemblages organize and regulate predatory functions within culture or between cultures. I apply the term "machinic" in reference to Deleuze and Guattari's (1987) notion of a posthuman machinic phylum in which abstract expression machines channel turbulent and self-organizing behavior in social systems that include matter-flow as well as sign-flow in the social process of subjectivization.

The sex-gender system once provided exemption for women from immediate participation in the state authorized killing-machine—from ritual sacrifice in ancient Aztec culture to military organizations in modernity—except of course as victims or support mechanisms (such as prostitution for "R&R," intelligence gathering, data processing, and computation, etc.). I am not suggesting, then, that women have not participated indirectly in the killing-machine throughout the centuries, through all the ways by which they provided component bodies for or complied with regimes of state power that depended upon regimes of violence in order to operate. Indeed, the first military computers in the late nineteenth century were rooms full of women operating calculators, carrying out the large-scale computations needed for ballistic analysis (De Landa, 41).

But well into the twentieth century, the predatory position itself was historically thought of as always ultimately male. During World War II, however, women began to be systematically incorporated within the war machine by U.S. military recruitment through the Women's Army Corps. China after the Maoist revolution also incorporated women into the military, even at the level of ground infantry, as did the Israeli state in the 1948 War of Independence and the Eritrea resistance in Ethiopia.

I begin then by noting that the emergence of a lesbian serial killing-machine is occurring precisely at a time when the state is in contention with women over abortion practices and combat status, while the military's advanced research groups are removing humans from the decision-making loop of advanced predatory weapons-systems. My opening premise here is that the instance of the lesbian serial killer is not an aberration of femininity, but that it signs a new symptomatology of "normative" femininity which we see emerging in postmodernity.

In an epoch marked by mass-produced feticide and a militarized feminine, it's not surprising that the cinematic screen machine is projecting a symptomatic nightmare image for public consumption: the woman who murders more than once, in cold blood. The semic code organizing this cultural figure is hardly new, however. The myth of Lizzie Borden provided a prototype of the contemporary feminine predator emerging within the popular image repertoire of late nineteenth-century America. While in actuality Borden was acquitted of murder charges (the case hinged upon the prosecution's inability to find the murder weapon and the appeal of prominent members of the community on Borden's behalf), the Borden myth both popularized the threat of a feminine-predatory subject and contained it within the rubric of familial crimes of hate (i.e., patricide and matricide) or within the rubric of an asystematic psychotic break within the individual (i.e., insanity).

While the Borden myth reminds us that the feminine predator is not a new figure in U.S. popular culture, it also highlights the signifying difference in 1990s iterations of this old trope of dangerous femininity—that difference being the association of the feminine predator with a manifest homoeroticism. It's my premise here, however, that the new queer mark on this old predatory body screens a repetition that is in actuality the more telling signifier in this story of social violence and gendered identity—a repetition that points us toward a contagion of social violences so vast that it is threatening, as it is always threatening, to exceed the constative rationalizations of our notion of abstract justice. This excess of justice, which is the same as saying justice's failure, in regard to the lesbian's public faciality symptomatizes a failure to reify as "just" what is blatantly a sacrificial exchange of social bodies and social violences in which the lesbian channels violence for the community. I want to return for a moment more to the history of this screen projection whose being has been sent, both to the lesbian "I" and to the Other, for the long duration of a tumultuous and violent century.

The mass-production of the myth of a female predator was delivered into the twentieth century by the technology and broadcast distribution of the popular fiction industry and later the cinema. By the 1920s, '30s and '40s, the hard-boiled detective stories and crime novels of the Hammett-Chandler-Spillane tradition had modernized and Americanized the genre of crime fiction made immensely popular by Sir Arthur Conan Doyle in the Sherlock Holmes series which spanned the turn of the century from the 1880s to the 1920s. One of Dashiell Hammett's trademark innovations on Doyle's classic formula in his Continental Op stories published in the '20s and '30s was the insertion into the narrative of a type of murderous female character whose capacity to kill precisely exceeded the discursive limits of familial hate crimes. That her predations, rendered in classics such as "The Girl with the Silver Eyes," exceeded those of familial crimes of hate, however, only suggested all the more that her figure, like Borden's, was signifying symptomatically the breakdown of a dysfunctioning nuclear family. Raymond Chandler in the '30s and '40s continued to develop this murderous feminine character type, placing her in a social landscape of crime and violence in which moral conclusions were rarely drawn and in which truth and justice were manifestly determined by money and power, epitomized in his first novel *The Big Sleep* (1939). In the American milieu of signs in which the wealth, status, and tradition of Sir Arthur Conan Doyle's social order were overturned by greed, economic inequities, organized crime, and everyday immorality, the dark woman of crime fiction was becoming unanchored from both the protection and morality of the domestic sphere. In the fiction of Mickey Spillane, the detective's aggressivity toward the dark lady of crime was becoming overtly murderous and violent. In *I the Jury* (1947), for example, the detective Mike Hammer realizes that the woman he has been erotically desiring is also the criminal he is pursuing. Rather than arresting her in the tradition of Hammett's Sam Spade, however, he empties his revolver into her crotch.

Throughout the 1930s and '40s, this figure of a female predator was further popularized for broadcast audiences in the immensely successful genre of *film noir,* where she reigned supreme on the cinema's ghostly projection screen as the deadly yet seductive *femme fatale.* Reproducing the destabilization of gendered signs and identities that this figure of dangerous femininity introduced into crime fiction, she typically appeared in classic *noir* at center screen only to become by film's end either the arrested body that would rationalize the police force (as in Roy Del Ruth's and John Huston's filmic versions of Hammett's *The Maltese Falcon* [1931, 1941]), or the reassimilated wife rationalizing patrilineal property rights and validating the marriage system as signifier of a healthy social body (as in Michael Curtiz's *Casablanca* [1942]).

In the late '50s and '60s, television's electronic medium would conduct the spectre of this feminine predator, assimilating *noir*'s claustrophobic *mise en scène* to the tv studio, and freeing the cinematic apparatus to produce a technologically innovative product in the '60s less suitable to televisual mimesis and appropriation. This new product was the panoramic, cinemascopic Western—a nostalgic projection monumental enough to screen the translocation of the U.S. frontier from North America to Southeast Asia. Meanwhile, the collapse of the Hollywood studio system in the '60s, assisted by tv's competition, pushed the filmic detective genre toward further disintegration as a containment mechanism for the very phantom projection it had mass-produced.

By 1974, the channeling of violence through the fatal woman's phantom projection virtually imploded on the silver screen in Polanski's *Chinatown,* a film in which the *femme fatale* is an incest victim who, though innocent of any crime, is shot in the eye at film's end by the police, while the evil father makes off with his daughter/granddaughter wrapped in his incestuous arms while she screams hysterically for her dead mommy. Against this historical backdrop of imprisoned and dead *femmes fatales,* transformations in '90s popular representations of women-who-kill appear not as a rupture or discontinuity, but as qualitative changes in the woman's relation to the predatory act, to the tools of violence and organs of predation, and to the ability of the modern legal system to successfully screen either feminine depredations or the constative production of women as sacrificial victims of social violence. In addition, the new movie genre coined by *Mirabella*'s June '92 issue as "psychofemme" films—a genre which includes *Thelma and Louise, Single White Female,* and *Basic Instinct* among many others—increasingly depicts a failure of the heterosexual economy to regulate and contain homoerotic exchanges between women.

In 1974, the popular narrative in *Chinatown* screened by Hollywood to a post-Civil Rights, post-Watergate and post-Vietnam public depicted the police and a patrilineal property system as the embodiments of, not the protections against, a systematic violence toward women. Nearly twenty years later, the '90s cinema has been busily (dis)organizing a cultural representation of an angry and vengeful woman—a woman who threatens to step beyond the bounds of heterosexual exchange—into two screen bodies, two clichéd images, which are two faces of the same fatal figure whose cold-blooded depredations threaten the nuclear family, the state police force, and heterosexual law and order. One of those screen bodies channels violence toward men and the nuclear family while escaping not only the economies of clan retribution and sacrificial substitution which the modern justice system supposedly displaced, but also the modern justice system itself. This screen woman is embodied paradigmatically in Sharon Stone's unpunished predations in *Basic Instinct* in 1992. Her body double, in contradistinction, channels violence toward men and the nuclear family, but is in turn subject to both a narrative economy of violent retribution and a symbolic economy within the public sphere of ritual sacrifice. The fate of the double is epitomized by Madonna's bloody, sacrificial expenditure at the end of *Body of Evidence* which appeared one year after *Basic Instinct*'s release.

Similarly, if *Basic Instinct* depicts a lesbian predator who kills yet eludes both retribution and justice, it's because her lesbian double in the film, Roxy, dies in her place, while Catherine Tremell succumbs to a heterosexual exchange from a homosexual plenitude. In spite of that exchange, however tentative if not volatile it is rendered by the film's doubled ending, the predatory screen face projected in *Basic Instinct* does not reassimilate into the nuclear family, nor does she become the target of violent retribution. And in a now common deviation from classic *noir* narratives, she also does not go to jail, eluding reappropriation and containment under a modern economy of abstract justice embodied by the state police and juridical system. It's hardly coincidental that this juridical system was revealed in the streets of L. A., in the same year as *Basic Instinct*'s release, to be no more than a veil for violent retribution and overt racism, and it was exposed as such by the unofficial video documenting Rodney King's beating at the hands of L. A. police which came to haunt the U.S. court system as well as the evening news.

It's not surprising then that the danger represented by Stone's predations, only tentatively contained by her marginally secured heterosexuality, is transducted into violence toward her body double, a woman who kills men and is violently killed by them in turn, embodying an exchange of violences not rationalized by a system of abstract justice as in classic *noir*, but by vengeful retribution on the level of the individual narrative and sacrificial substitution on the level of an intertextual public sphere. In other words, if Sharon Stone can walk after both lesbianism and serial murder in *Basic Instinct,* a sacrificial screen body no less culturally loaded than a post-*Sex,* bisexual Madonna must be offered up for a mass act of ritual murder, partly played in slow motion with background music, in *Body of Evidence*'s normalizing repetition of *Instinct*'s aberrations against the legitimate exchange of social bodies and social violences. The most symptomatic difference, then, in '90s screenings of this old threat of feminine violence lies in the now doubled representation of the law's inability to regulate either the social violence channeled toward this woman or the violence that she channels toward the socius. Both bodies expose a system of violent exchanges once screened by the communally held and officially sanctioned pretense of an abstract and therefore just equivalency under the law. Whether the woman walks after lesbianism and serial murder or whether she is blown away in an unveiled spasm of violent retribution in the spirit of *Fatal Attraction*'s (1987) vulgar, public blood-letting, 1990s filmic representations of feminine predators figure a woman who is beyond both the punishment *and* protection of the justice system and its police force—beyond precisely that which was challenged but always finally rationalized and legitimized in classic *noir.*

Obviously, the lesbian and gay community needs to do just violence to this screen projection of a lesbian psychofemme predator, but in doing so, it also needs to address the more broad and difficult issue of the proper collective relation to a contagion of social violences. That is, the nightmare projection of a lesbian predator directs us to take up the more general question of how to represent, to ourselves and to the other, violence's past and the past of violence in the lesbian community. Because beyond that question, an even more pressing issue faces us—the issue of violences and resistances in our own discourses about collective identities and public representations. Taking this broad approach to the relation of lesbians and social violence within the general economy will require several detours that we will later find are only rerouting opportunities for the essential adestination of lesbian identities who have sent their being and who have had their being sent into the collision course of becoming-woman and becoming-killing-machine.

In the first week of January of 1993, the *New York Times* reported that AIDS had become the leading cause of death in the New York state prison system, while the U.S. military had renewed "low intensity" bombing raids in Iraq to stabilize "global security." Later that same month, the Joint Chiefs of Staff announced to the public their resistance to Clinton's campaign promise to end the 1948 ban of gay men and lesbians in the military, citing the violent beating-murder of a gay sailor by his mates as an example of the "disciplinary" problems that could arise from revoking the ban. By the end of January, three marines in Wilmington, North Carolina, pulled a patron out of a gay bar and fractured his skull, shouting "Clinton must pay" (*New York Times,* February 2, 1993). The following week in the Bronx, a young African American man suffering from seizures and a drug problem, who had been in and out of shelters, prisons and emergency rooms for years, beat an 80-year old woman to death with a lead pipe outside her church (*New York Times,* February 7, 1993). Five months earlier in Cleveland, a lesbian had murdered her lover, saying that the lover's father had harassed his daughter for being a lesbian to the point that she announced, "I wish I was dead" (Pontoni). In his study of the systematization of violence in the *The Nervous System* (1992), Mick Taussig recounts the story of a woman who, when approached by a social worker who has arrived to take her last remaining child as a ward of the state, pulls down her pants and reveals a stab wound in her butt. Upon realizing what she'd done, she pulls up her pants and laughs nervously.

What I'm suggesting is that we live in an epoch of continuous violences, and in this our epoch differs little from previous ones, except perhaps that today the right to participate in military predatory systems has come to signify the full social status and civil rights of the individual citizen, whether that citizen be lesbian or gay, a woman, or African American. It's from this epochal perspective of the general economy that I seek the proper affect with which to approach the lesbian body which Sharon Stone's projection in *Basic Instinct* screens as body double. That body, which could never appear unscreened in the public sphere, belongs to Aileen Wuornos who, one year before *Basic Instinct*'s release, was arrested for murdering seven men on seven separate occasions along a rural highway in south Florida, and who was labeled by the media as "the first lesbian serial killer."

Bearing Wuornos's body and public sign in mind, I want you to recall the mediated outrage from the lesbian and gay community, vocalized by Queer Nation, Out in Film, and the national Gay and Lesbian Alliance Against Defamation (Glaad) in regard to *Basic Instinct*'s 1992 portrayal of the lesbian as serial killer (*New York Times,* March 29, 1992). We all know that gays and lesbians as liminal and marginalized social bodies are particularly susceptible to the channeling of social violence, as are other minority bodies. The proportion of *psychofemme* lesbians to appear on the silver screen compared to the total number of lesbians in general to appear on screen speaks for itself as to the state of just and equivalent representation in the mainstream cinema, and epitomizes the use of the lesbian body to channel and then screen a potential contagion of violence erupting from the breakdown of the sex-gender system in the so-called "healthy" heterosexual social body. But in order to understand the systematicity of violence within the killing-machine—what Taussig (1992) terms "terror as usual" in the Nervous System—and in order to understand the roles that lesbian bodies play within such a system, we must go further than the just and timely protests by Glaad, Queer Nation, and Out in Film, and be willing to push on to question the politics of violence and resistance in the stories we tell *about* violence in the lesbian community as well as in the mainstream media, in both the official and unofficial representations of violence, and in our personal and public representations of violence (38).

For a few moments, then, I want to pretend we don't have to tell this official and utopic story anymore in which lesbians like U.S. citizens in general and like women in general are non-violent social subjects, in an argument which attempts to protect lesbians from the violence of social defamation and the physical and emotional brutality that goes with it by claiming their own innocence from participation in violent systems of social exchange. This claim touches us all the more painfully when we allow ourselves to know what is everywhere obvious, that all systems of social exchange known to men and women in our culture depend upon systematic exchanges of violence. Some of those violences are sacred, legitimate, and public; others are criminal, illegitimate, and covert. And the regulation and breakdown of that distinction is precisely the domain of the killing-machine.

Let's make no mistake that the unifying factor in the disparate group of "legitimate" victims of violence is their liminal status on the fringes of society—the degree of their dis-integration within the legitimating discourses of the healthy socius (Girard). Because of their marginalized status, the exposure of these social bodies to violence entails little risk of reprisal from any empowered social group. The constative exchange structuring the sacrifice is thus an act of violence without risk of vengeance. It's within this economy of exchanges which the modern justice system supposedly replaced that we can begin to understand the meaning of the lesbian serial killer's serialized and mediated signature. Once protected from sacrificial substitution because of risk of clan retaliation, women become subject to increasing social violence with the breakdown first of local communities and later of the nuclear family. As the social link between women and the larger community disintegrates, women, particularly single women or women outside of the economy of heterosexual exchange, become increasingly marked for violence. The serial killing-machine typically preys on women who were once protected by the status of the local clan as it was reproduced in women's maternal function. Thus, single women are particularly susceptible to serial killing, as exemplified by the Green Rivers slayings of prostitutes in Washington state, Ted Bundy's interstate murder spree of young single white women, and the Gainesville murders of white coeds at the University of Florida. Deterritorializing local cultural geographies of class and race in order to reterritorialize them under a regime of terror, the serial killer is typically itinerant, often using transportation technologies to better prey on the disintegrating social ties of broken-down communities in which neighbors don't visually recognize or speak to neighbors and the comings and goings of "strangers" can no longer be policed by communal social bonds.

In the late modern cinema's projection of a lesbian predator, the failure of the clan system to regulate social violence toward women takes a new turn. In *Single White Female* (1992) the murders are committed specifically to spite the clan, as well as the feminine positions of 'het' daughter and wife that the kinship system authorizes. In *Basic Instinct,* the murders are against the clan. If nothing else, *Basic Instinct* projects the breakdown of the nuclear family system: mothers who kill their children, daughters who kill their parents, girlfriends who murder lovers, girlfriends who are lesbians, female psychotherapists who hate their male patients, women who hate the idea of domesticity and child rearing, phallic women who expose gaps in the law, *femmes fatale* who murder the detective. This projected nightmare drives the fantasmatic narrative to its nearly classical resolution—the detective murders the murderous lesbian double, becomes the object of the *femme fatale*'s desire, and re-establishes phallic order (. . . almost).

In the case of Wuornos, the murders are both against the clan and in and of themselves the revenge of the clan which she has taken upon herself to execute—that is, they stand in the place of an absent clan revenge while they also stand in judgment against domestic violence within the clan. In court testimony, Wuornos maintained that she killed only those johns who "deserved it." Aileen Wuornos, as a homeless lesbian highway prostitute, is the liminal body that should have constituted the body of the sacrificial victim. Refusing that role, she chooses instead, illegitimately, the position of the sacrificial priest, the judge, and the executioner—until she is caught by the police and brought under the judication of the legal system, where the power of the act of violence which she deterritorialized for her own signature is reterritorialized by the Law and redirected toward Wuornos as a proper body for social violence in the form of the death sentence.

Serial killing treads the liminal social space between sacrifice and murder—between the impure sacred and the criminal. It is a form of anonymous violence that foils retribution from a community, becoming then a game of outwitting the state police force. It exists in the space between the emptied function of the local community and family clan in the late modern state and the state and national judicial system. Its intercounty or interstate flight is countered only by an intrastate and interstate police data system. In this sense, serial killing is a game played with an abstract machine (deductive rational logic authorized as the law and institutionalized as a computerized police surveillance and data-processing system) in which the serial killer tracks the disintegrating or missing social links through which the body politic reproduces and rationalizes power, rather than an exchange executed through a body within the context of a specific community.

In an economy of sacrifice, the desire to commit acts of violence toward those near us is channeled to bodies on the margins of society, thereby quelling a potential contagion of violences within the legitimate socius. Serial killing spurs rather than quells a contagion of violence the more it takes inappropriate and illegitimate victims as its target—white coeds as the privileged daughters of the rising bourgeois middle class or *nouveau riche* in the case of the Gainesville murders, for example, or, in the case of Wuornos, white heterosexual men. It's not surprising that this contagion of violence is then redirected toward "appropriate" victims, just as south Florida newspapers reported that hate crimes against lesbians increased during the period of Wuornos's trial (Brownworth).

The machinically assembled "seriature" of the lesbian serial killer—a term coined by Ronell in her reading of Rodney King's assembled media representation in "Haunted TV" (1992)—is the effect of a network of media representations. This network constituting Wuornos's public signature as serial killer—her seriature—includes Hollywood projections of women who kill and media coverage of Wuornos as the "first lesbian serial killer" in the form of news reports, coverage of the trial on "Court TV," and the made-for-tv movie *Overkill.* This network produces mediating windows through which the socius polices itself through the perpetual threat of pandemonium from an imaginary exterior. The temporal flow of rational consciousness in tv then frames, formats, and fractures this pandemonium which it constitutes. In the process, the face of the serial killer can only sign in the public sphere as manifestly posthuman, in that it signs as the concrete form of expression of an abstract machine mediating social violences. In this mediation which now constitutes the public sphere, the post-Cartesian body signs as "I am because I appear on network news."

In the made-for-tv movie depicting the Wuornos story, for example, television simulated its own call to ethics as realistic docudrama. That which was usually hidden from view suddenly burst in mimetic packaging replete with interrupting commercials onto the broadcast screen of evening tv—as the ugly scene of social violence embodied in the "robbing" of a street prostitute (i.e., rape) and, even more censored because far more violently contagious for the community—*the revenge of the robbed prostitute* who is both a lesbian and an incest "survivor." *Overkill* signifies first and foremost revenge rather than self-defense. Aileen Wuornos as "seriated" killer serialized by her mediated signature is the liminal scapegoat who refuses to become the sacrificial body she is marked to be by channeling social violence rather than absorbing it, by refusing to eat the poisonous pharmakon (first as the phallus and penis of the incestuous grandfather and uncle, later as the hard-on of the abusive and robbing trick) by which she would be transformed into the sacrificial body as lower-class, lesbian, second-generation incest victim and streetwalker whose tricks often enough don't even pay her (i.e., in street slang they rob her). Instead of eating this poison of the socius, she *becomes* the poison that must in turn be eaten by the sick community in a blatantly sacrificial exchange reified as an economy of abstract justice, in order for that economy and that community to be cured of the outcomes of its own violence—that is, in order to represent to itself the normative "health" of its everyday social relations. This is the circuit of mediated exchanges into which Aileen Wuornos sends her identity and has her identity sent as lesbian serial killer.

Wuornos's violence is the effect and the affect of an illegitimate social identity that can only become legitimate as serial killer. In Eve Sedgwick's terms laid out in an article on queer performativity (1993), the extent to which Wuornos doesn't have a legitimate identity is the extent to which she has an identity of shame. Her performance of her shame, and shamelessness, is the only way she has to legitimate—that is, to make visible and intelligible to herself and to the other—the affect of anger which constitutes her identity as one who is unable to *be* as legitimate identity, but who refuses either to annihilate herself through suicide or to be the murdered body in the woods.

Her seriature is reterritorialized, assembled and regulated by an interstate police, computer database, broadcast media network. Even as she threatens the legitimacy of that network, she evokes the "justified" use of police force, thereby providing the body upon which the social system's unjust violences can be reified as abstract justice. She both threatens to rupture the illusion of non-violent social relations within the body politic and provides the individuated body through which the body politic can represent itself as non-violent and just. Her telemediated identity is framed and serialized just as her telemediated trial was being serialized on "Court TV." For the television viewer of *Overkill* (which received vociferous mail complaining that Wuornos was represented too sympathetically), there is an uncanny recognition of the familiar in Wuornos's bizarre docudrama—her life *is* docudrama, before the docudrama of her life is ever produced. But more than that, her docudrama is completely mundane and predictable, her story makes the extraordinary (a woman's multiple murders of men) completely ordinary—a homeless prostitute without a pimp walking the edge, she found it more rewarding to turn tricks than be a motel maid, and later found it more lucrative to murder and rob bad johns than to fuck them for $20 or to let them rob or murder *her*. Meanwhile the legal system and the deductive rationalizations of popular crime-solving want to make the ordinary fragments of Wuornos's dismembered life into an extraordinary, exotic, and exemplary narrative of crime and punishment.

But what are we re-membering in the story of Aileen Wuornos? Is it that lesbian bodies have real relations to specific forms of social violence? Is it that sometimes it's difficult to tell the police from the bad johns, the honest men from the dishonest men, problematized in the made-for-tv movie by the undercover cop who both befriends Wuornos and betrays her, or by the dead johns who turn out to be a minister and a state social worker? Is it that Wuornos is the embodiment of social violence—as homeless woman, incest "survivor," rape and robbery victim, highway prostitute, and "killer dyke"? Is it the symptomatology of post-incest "profile" in her story: suicide attempt, self-hatred, anger, repression, multiple personalities, loss of family and the loss of economic stability that accompanies loss of family?

Are we re-membering the primal scene of the bad john being shot in the heart, brain, and genitals by a highway prostitute that he tried to "rob"—killed in the Florida woods already signed by the deracination of the Seminoles, by the economically condoned social violence against migrant labor, and by the overt oppression of poverty stricken African Americans, Hispanic Americans, Mexicans, Haitians, Jamaicans, etc.? Is it that Wuornos is the blonde white woman, the other other of *Paris Is Burning* (1991), poor white trash killing the offending family members, the incestuous grandfather and uncle, over and over and over? Is it the signature of repetition and death drive, not finally interiorized as self-mutilation and suicide so common to femininity's contemporary signature, but exteriorized as the seriated murder of the other?

Are we re-membering the coldness and cruelty she learned from the absent mother who abandoned her and who was also a victim of domestic violence, and from the father imprisoned for child molestation who hanged himself in prison—two generations of angry knowledge passed down as a sociopsychological and physiological body of signs, body memories, and night terrors. Is it that popular images of "killer dykes" provide a convenient screen memory for the massive workings of terror and violence of the killing-machine? Or is it that Wuornos, signing herself as the daughter at the bottom of the bottom of the white social hierarchy, is the last whipping post, the place where the buck stops, finally, as a storm of meaningless violences, flights of desperation, alcohol, outrage, alienation, and schizonoia—the last resort of illegitimate identity, like the bar bearing that name where Wuornos was finally arrested (MacNamara).

As homeless lesbian prostitute, Wuornos is beyond the protection of the nuclear family and the law. Without men to threaten retribution, she takes revenge into her own hands, but in doing so she is marked always and already as the victim, the scapegoat, and the sacrificial body. Theoretically, a just legal system can fill the gap of a missing family narrative, but we know, as does she, that that belief is only an abstraction. In actuality, Wuornos is alone in the woods with her johns beyond the protection of an abstract law that can only sign itself through the criminalization of her social body, and hers alone, not her grandfather's, not the bodies of her abusive tricks, and likewise, she for her part can only be known as a visible, intelligible body within the public sphere by the criminalization of her seriature as not suicide and not sacrifice. If she had been the streetwalker murdered in the woods by one of her tricks or if her early suicide attempt had been successful, Wuornos's signature would have remained forever invisible to the public sphere which, in theory, bears the responsibility for mitigating the violence of inequivalencies within civil society and guaranteeing abstract justice for all (Habermas).

If the difference of Wuornos's sacrificial sign becomes her lesbianism in the popular imaginary (which it has), then the contagion of violence will spread (which it is, increasing the level of violences channeled onto the lesbian social body at large). But let us not forget that the real difference driving Wuornos's life story is her poverty, her economic oppression, her alienation through loss of family typical of adult women subjected to incest as children, her liminal social status as homeless highway prostitute—all that which is uncannily familiar in her story, mundane, completely ordinary, and understandable, that is, her lifetime exposure to unadulterated forms of "pure" social violence. That she has a lesbian lover is the one tender spot in a horrible and sordid sequence of events that constitute the limit-text of her life story. It is the one amorous act of volition that stands counter to her only other act of will that can constitute a signature of her agency as one who refuses victimization—the murder of the other who would Other her. As such, her lesbianism is not the key but the tangential detail of the seriated narrative that must be exoticized by the mass-mediated killing-machine, that is, made extraordinary, but only as cliché, even as its real difference is exorcised, screened, and erased.

In the end, expertly worked over by the state police force and under threat of criminal charges, the lover Tyria Moore is led to testify against Wuornos in court—the final betrayal in a long list of betrayals by those who were supposed to "love" her—and the successful rupture of any potential collective lesbian social identity. Understanding the mechanism of Wuornos's lifelong alienation—the systematic reification of all her social relations into relations of violence authorized by regimes of paternal property or commodity exchange on one hand, and the erasure of her illegitimate and unofficial but self-determined relations on the other—Arlene Pralle, a "stranger" who like Wuornos was adopted as an infant, states to the press, "there but by the grace of God I would be," and adopts Wuornos after her imprisonment, thereby inscribing in the public sphere for Wuornos, and all her doubles, at least the signifier of a tangible, permanent, and self-determined social relation that the Law can imprison, but cannot, finally, erase.

Meanwhile, Wuornos's seriature—at the juncture of becoming-lesbian and becoming killing-machine constituting the criminalization of lesbian identity—is time-coded for a feature film, for a few more headlines, for another made-for-tv movie or independent documentary perhaps, before the chair ends her narrative according to the script she herself wants to write—with a jolt of pure electricity—like the electric circuits of television's simulation of Wuornos's life story. That story is one that she never really had, until tv made it up by giving it at the same time broadcast life, temporal frame, and narrative closure—seriating fragments of a life into mainstream tv that will never add up to either a meaningful story or a just ajudication.

Works Cited

Brownworth, Victoria. "Killer Lesbians." *Village Voice*, vol. 37, no. 4, October 13, 1992: 25.

Chandler, Raymond. *The Big Sleep*. New York: Ballantine Books, 1971.

De Landa, Manuel. *War in the Age of Intelligent Machines*. New York: Zone Books, 1991.

Deleuze, Gilles, and Felix Guattari. *A Thousand Plateaus: Capitalism and Schizophrenia*. Trans. Brian Massumi. Minneapolis: University of Minnesota Press, 1987.

Girard, René. *Violence and the Sacred*. Trans. Patrick Gregory. Baltimore: Johns Hopkins University Press, 1977.

Habermas, Jürgen. *The Structural Transformation of the Public Sphere: An Inquiry into a Category of Bourgeois Society*. Trans. Thomas Burger. Cambridge, MA: MIT Press, 1992.

Hammett, Dashiell. *The Continental Op.* New York: Vintage Books, 1975.
MacNamara, Mark. "The Kiss-And-Kill Spree." *Vanity Fair,* September 1991: 90–106.
Pontoni, Martha. "Woman Kills Lover." *Gay People's Chronicle,* Cleveland. Vol. 8, issue 2, August 21, 1992: 1.
Ronell, Avital. "Haunted TV: Rodney King/Video/Trauma." *Artforum,* September 1992: 71–74.
Sedgwick, Eve. "Queer Performativity: Henry James' *The Art of the Novel.*" *Gay/Lesbian Quarterly,* vol. 1, no. 1, 1993.
Spillane, Mickey. *I the Jury.* New York: New American Library, 1954.
Taussig, Michael. *The Nervous System.* New York: Routledge, 1992.

NINE

"Death of the Family," or, Keeping Human Beings Human

Roddey Reid

> The family is the most human, the most powerful, and by far the most economical system for making and *keeping human beings human*. . . . Throughout modern industrial societies and across diverse cultures, the dominant and most successful family form has been the two-parent family. . . . Indeed, anthropologists long have recognized the two-parent home . . . as a common foundation of human societies.
>
> —*Families First.* Report of the National Commission on America's Urban Families (1993)

> The dirty little secret in place of the wide open spaces glimpsed for a moment. *The familialist reduction, in place of the drift of desire.* In place of the great decoded flows, little streams recoded in mommy's bed. Interiority in place of a new relationship with the outside.
>
> —Deleuze and Guattari, *Anti-Oedipus*

Human Interest Stories

The melodramatic title to my essay is a play on the din that assaults ear and eye as we punch through the digital presets and surf on remote control at home or flip through magazines in the doctor's office anxiously awaiting the latest report on what's transpiring on and below our skins. Dramas proximate and far away swirl around and past us telling tales of "family" under threat: from murder-suicides of family members, crack babies, day-care child molesters, the HIV virus, and white Yuppie mothers hiring undeclared Latina nannies to parents stealing away from their children for a quick Mexican vacation, vicious baby-sitters, bisexual husbands, mothers who smoke during and after pregnancy, and deadbeat dads. These stories keep you on edge anxiously looking over your shoulder at one moment and at another peering into the eyes of friend, neighbor, spouse, or lover, wondering: *who's next*?

Or rather that question *should* cross your mind if not your lips, for don't these stories raise your pulse rate, get you in a cold sweat, have you call your city council or U.S. senator? If not, well then, there's clearly something wrong with *you*. You're devoid of feeling, you don't care, you don't belong to The Community. For it's time to defend The Family against all comers and all enemies.

The stakes are high, for not only *our very humanity* is at risk, as suggested by the first epigraph, but also the very future of the nation. The "Family Strength Index" is at an all-time low, warns the same government report published in 1993:

> The family trend of our time is the disinstitutionalization of marriage and the steady disintegration of the mother-father child raising unit. This trend of family fragmentation is reflected primarily in the high rate of divorce among parents and the growing prevalence of parents who do not marry. No domestic trend is more threatening to the well-being of our children and to long-term national security. (*Families First:* 19)[1]

"Family" is the alpha and omega of national life. As George Bush put it in a commencement address in 1992, "Whatever form our most pressing problems take, all are related to the disintegration of the American family" (*Families First:* 13).

Night of the Living Dead

It's everywhere and it's catching. Families and households are dead and dying all around us, we are told. In the gothic version of the tale, families are dead but never quite: the undead. Zombie-like, they wander across tv screens, billboards, and newspapers, displaying their gaping wounds and rotting flesh and set to feed on our peace of mind and inner tranquillity. Once they get to you, you too will toss and turn at night and begin your insomniac career through public and private space in search of new bodies and minds to nourish the Great Fear. Danger lurks in every corner, at home, across town, and abroad. How can you sleep when you watch a docudrama that reveals that California's Family of the Year in 1965 was a hotbed of child abuse? ("Shattered Trust" 1993). Or that for one kid in fifty, one-third of that sacred triad Mommy-Daddy-Me may be in jail at any one time? (Coleman 1993). And what's left of your peace of mind once you've read the remarks of Dawn Stover, the associate director of the Portland, Oregon, group called Advocates for Life, who tallies up the score in the abortion battle this way: " 'There are 30 million dead children, one dead abortionist and one wounded,' Stover said. 'Who's winning the battle? Not the preborn who've not yet breathed their first breath of life' " ("Pleas for Protection" 1993). Or, finally, when the headline jumps out at you that a prosperous Los Angeles businessman was shot dead beside his mother's grave, which he visited every week, while kneeling and holding a dozen roses (Corwin 1993). Nothing is sacred anymore.

Save the Children

If we all shuddered on the eve of the Gulf War at the sight of that mother of all child molesters, Saddam Hussein, caressing the heads of captive American children on CNN, at least the allies brought the hostages home and took care of *him*. On the home front, things are equally worrisome for children if not more so. Who can you entrust care of your children to outside the home? Day-care centers teem with "child predators" who seduce and abuse their young charges. Something must be done. Just as former Secretary of Education and Drug Czar William J. Bennett claimed that "we must speak up for the family. We have to say it, we have to say it loudly, we have to say it over and over again" (*The Family* 1987: 24), so too we must speak up for children, for the bodies and speech of the little ones are no less fragile and endangered than The Family. They are *in-fans* by nature: without (credible) voice. Or so declared an 11-year-old girl who testified at both of the McMartin day-care center trials in Los Angeles, which ended in deadlocked juries seven years after the first charges were filed against Ray Buckey, setting off a wave of hysteria in Manhattan Beach and surrounding communities. There were initially

> allegations not only of rape, sodomy, oral copulation, and other sex crimes, but also of pornographic photography sessions, "naked games," field trips away from the school for illicit purposes, animal mutilation, threats and satanic-like ritual and sacrifice. (Timnick 1990: A1).

After the second mistrial, the young witness was

> disappointed and really angry. I thought they'd believe me and say he's guilty. I sort of think it's because we're kids, and people think all kids lie. . . . But he deserves to be in jail. He knows what he did. He knows he's a bad person. (Timnick 1990: A1)

Periodically, from 1983 to 1990 the Manhattan Beach parents who claimed their children had been abused held banners outside the courtroom that read, "We Believe the Children" but the juries did not. Who is going to tell the children's tale? You? Me? The Children's Institute of Los Angeles?

Panicked parents of Edenton, North Carolina, who banded together as Citizens Against Child Abuse, fared better. The case was featured in an ABC "Frontline" documentary as well as in another aired in July 1993 by PBS titled, "Innocence Lost, the Verdict." The jury convicted Robert F. Kelly, co-owner of Little Rascals Day-Care Center, of 97 counts of abusing 12 children, and he was sentenced to 12 consecutive life terms. " 'The children were convincing,' " said a juror, much to the parents' relief (Smothers 1992a: A18 and 1992b: A16). However, these days the anxiety over abused children is such that the "speaker's benefit," as Foucault once put it (1980: 113), of speaking for the silenced isn't good enough; greater voice and authority can be had by claiming to *be* one of the voiceless.

Now few can claim the status of a fetus, even fewer that of an aborted fetus, whereas all of us may have been abused during childhood. This may account in part for the new wave of hysteria around child abuse. The appeal of that subject position is so irresistible that during jury deliberations one juror in the Kelly case melodramatically "revealed" to have been abused as a child (Smothers 1993a: B9). But he was not alone. Oprah Winfrey, the national talk show host never known to be behind the times, had already preceded him. She had confessed on her nationally televised show that she had suffered repeated abuse as a child and had launched a national crusade that took her before the Senate Judiciary Committee in November 1991. To the assembled committee she proposed a National Child Protection Act drawn up for her by a hot-shot Chicago law firm she had hired declaring, " 'I am committed to using all my will to follow through on this legislation, and on the issue of child abuse. . . . I intend to make this my second career' " (Mills 1991: A1). A year later her crusade led her to narrate a documentary, "Scared Silent: Exposing and Ending Child Abuse," broadcast simultaneously on three national tv networks in August 1992 (Carter 1992: C13). The Abused Child has become shameless wannabes' object of choice, though those family members suffering from multiple personality disorders (MPDs) or victimized by "secondary smoke" are not far behind. The speech of victimhood (and its benefits) is one that won't be denied.

Toxic Parents

Oprah's campaign sounds a wake-up call to America: like communists during the Cold War, the enemies are not simply without the family but also *within.* Child abuse is a family affair, too, and may even push children to the ultimate crime: parricide. Witness the story of two brothers twelve and fifteen in Rush Springs, Oklahoma, who killed their father with a rifle while he slept, accusing him of repeatedly beating them and of molesting their ten-year-old sister. Abuse seems to know no limits: not only were journalists shocked by reports of physical and sexual abuse but they were almost as scandalized by the fact that he had deprived them of our national consumer culture by confining them to home except for school: we learn that "as police officers drove the boys to Lawston, where they are confined in a juvenile detention center, they stopped at a McDonald's for a meal. The boys had never eaten one" (Verhovek 1993: 1–9).[2] The father's crimes were such that the good folk of Rush Springs, which claims to be the watermelon capital of the world, rallied around the boys and desired nothing better than to forgive and forget and return to normalcy. " 'That's what we need to focus our attention on now, is watermelons,' said Justin Jones, a watermelon farmer" (Verhovek 1993: 1–9). Even families in the best neighborhoods may harbor the worst: David and Sharon Schoo of an upper-middle-class neighborhood of a town outside Chicago abandoned their two small girls for a week's vacation in Acapulco over Christmas and upon their return were arrested and released only after posting a $50,000 bond ("Leaving Two Home Alone"). Loving parents are no guarantee of safety: Teddy Lee Prichard of Wayne, West Virginia, took the lives of his wife and two of three children

> because he loved them and didn't want a court order to break the family up, relatives said. . . . "He loved his wife and kids so much that he killed them, if you can make sense of that. I am a sane person. I can't," said Prichard's brother-in-law, Evrett Hager. ("Dad Kills Wife")

Fathers threaten their families' life and limb, especially if they are smokers. One anti-smoking campaign billboard in Southern California features the face of an African-American man with a lit cigarette between his lips and reads, "Eric Jones put out a contract on his family for $2.65." And mothers can't be trusted with the sacred task of nourishing their children. One mother was recently sent to jail for tampering with baby food ("Mother Goes to Jail"). Toxic parents indeed.[3]

Something is always off in the family equation; if it's not the parents, then it's the children. For example, your kid may be suffering from that newest clinical entity, "teenage syndrome," e.g., "behaviors that tend to hang together (smoking, drinking, early and frequent sexual experiences, and, in the more extreme cases, drugs, suicide, vandalism, violence, and criminal acts)" (*Families First:* 15). Or, unbeknownst to you, your kid may turn out to be a closeted fascist. When Jonathan Preston Haynes's father heard allegations that his son, who had confessed to murdering a San Francisco make-over artist and a Chicago plastic surgeon for giving people "fake Aryan cosmetics," was a neo-Nazi, he exclaimed:

"I'm his father, and I never saw a clue of that," he said. Likewise the father said he had not considered Haynes to be mentally ill.

"He went ahead and got himself a degree and a series of good jobs, and there was every indication of smooth sailing," the father said. "We knew we had an unusual boy, but we also knew he was a very intelligent boy. There are enough eccentrics around Northern California (that) to have one in your family doesn't normally shake you up alot." (Williams 1993: A1)

Return of the Vanished Archaic Despot

With danger at every turn in family life, perhaps some return to the old patriarchal certainties of yore is in order. That is what the courts seem to have in mind, if decisions regarding Baby Jessica are any indication: she was forcibly handed over to her biological father after spending the first two and one-half years of her life with a couple (Cowley 1993: 54). Still, there was demonstrable public uneasiness about the whole affair, and certainly other events have not proved reassuring: the same issue of *Newsweek* that broke the Baby Jessica saga, "Standing Up for Fathers," featured as its cover story "Death Wish" with an enlarged portrait of David Koresh engulfed in the flames that destroyed the Branch Davidian compound in Waco, Texas, killing 85 people ("Death Wish"). Patriarchal perfection has its risks. As Robyn Bunds, a former cult member who was one of Koresh's wives from age 14 to 17, explained to *Newsweek,* " 'He's perfection, and he's going to father your children. What more can you ask for?' In fact," continues *Newsweek,* "Bunds says she was so committed to Koresh that she left in 1990, nine months after Koresh started sleeping with her mother, because she was tired of the abuse" ("Children of the Cult": 50). In the Waco confrontation both sides had one thing in common: they justified their actions through the trope of "family." In a televised press conference Bill Clinton defended the decision to have Federal agents move in on the Davidian compound in terms of the children at risk. He related a conversation he had with Attorney General Janet Reno:

> But in the end, the last comment I had from Janet Reno is when—and I talked to her on Sunday—I said, "Now I want you to tell me once more why you believe, not why they believe, why you believe we should move now rather than wait some more." And she said, "It's because of the children. They have evidence that those children are still being abused and that they're in increasingly unsafe conditions, and that they don't think it will get any easier with the passage of

time. You have to take their word for that. So that is where I think things stand." (Clinton 1993)

Meanwhile, Dick Guerin, the lawyer for Koresh's mother, complained that the government's actions had violated the sanctity of family life and the right to privacy: " 'At what point does society have a right to step in and say you have to raise your family our way? It's applying Yuppie values to people who choose to live differently' " ("Children of the Cult": 50). Anything can happen as "family" or in its name: recently, Argentinean police raided a sex cult which held 268 children. What did it call itself? "The Family" (Nash 1993).

Beyond Rhetoric: Scholars, Experts, and the New Consensus

These human interest stories writ large as national or even international catastrophes aren't thrown up by popular culture and the mass media alone; they are produced by an endless stream of researchers, expert witnesses, professors, and government officials who profess alarm at the imminent collapse of the American Family and with it the entire body politic. Government publications from the Reagan White House's *The Family: Preserving America's Future* (1987) to the bipartisan report of the National Commission on Children, *Beyond Rhetoric: A New American Agenda for Children and Families* (1991), and the report of the National Commission on America's Urban Families, *Families First* (1993) have joined new and not so new publications put out by political think tanks and academic research councils: from *Putting Children First: A Progressive Family Policy for the 1990s* by the Democratic Leadership Council's Progressive Policy Institute, *Free to Be Family* (1992) by the Reaganite Family Research Council to the national platform of the Communitarian Network and the work of the academic Council of Families in America, from David Popenoe's *Disturbing the Nest: Family Change and Decline in Modern Societies* (1988) and Sar A. Levitan, Richard S. Belous, and Frank Gallo's *What's Happening to the American Family?* (1988) to the anthology *Rebuilding the Nest: A New Commitment to the American Family* (1990) edited by David Blankenhorn, founder and president of the Institute for American Values.

All of these studies pack a one-two punch; they throw you off balance with frightening tales of national decline on the one hand and extend reassurances on the other that *you are not alone,* that there's a new meeting of minds between left and right at hand. There is Hope. Writes neo-liberal David Popenoe, referring to *Families First* and *Beyond Rhetoric:* "During the past several years, a fragile, but important bipartisan consensus has emerged among both policy makers and scholars on the central challenges of the American family" (1993). He echoes directly the claims of the largely neo-conservative report, *Families First*:

> Witness after witness—regardless of race, gender, marital, or economic status, or partisan affiliation—told the Commission that fragmentation of families poses a threat to the nation. Liberal and conservative scholars agree that the strength of the nation's family life is indicative of the nation's health and well-being. (*Families First:* 12)

The New Consensus is of course beyond rhetoric, thus beyond ideology, yet still firmly if gently normative, based on expert social science:

> Clearly, then, quite apart from political affiliation or ideological sympathy, current scholarship and bipartisan findings strongly confirm the importance of the intact, two-parent home to children and to the society as a whole. This finding does not, of course, disparage or condemn those who find themselves in other family arrangements, but is instead based on careful consideration of social science research and the testimony of witnesses regarding the benefits of the two-parent home. (*Families First:* 18)

Of course.

The startling political possibilities afforded by the New Consensus were made clear for all to see after Oprah Winfrey's testimony to the Senate Judiciary Committee, when she emerged accompanied by Strom Thurmond, one of the staunchest southern conservatives. According to the *New York Times,* "Strom Thurmond, South Carolina Republican and senior member of the committee later introduced Winfrey to the press, holding her hand, calling her a great woman" (Mills 1991: B1). *E pluribus unum.* The refrain returns over and over again that "family" unites us all and sets aside divisions of race and class (and presumably, gender). The authors of *Families First* claim:

> Nor is the family debate a coded or indirect way of debating issues such as race or poverty. It is not the same as debating the proper size of government or debating liberalism versus conservatism, Republican versus Democrat. It is about the health of the nation's primary social institution. (*Families First:* 11)

All of "us" are affected: "Family disintegration affects all Americans, from urban to rural, from poor to rich. As one witness told the Commission,

'No family is exempt from the kinds of problems that are going on in this country right now' " (*Families First:* 11).

"Us" and "Them": Writing the Social Body through "Family"

So the din is deafening. Families are dead and dying all around us goes the chorus.[4] Of course, some family households are deader than others. Dan Quayle and other entitled white men and women before him have made sure we understand that. All those households headed by women in the African American community, those Black Murphy Browns, now there's the cause of the 1992 Los Angeles riots ("Quayle Deplores").[5] And when Federal Judge John G. Davies handed down light sentences in August 1993 to LAPD officers Powell and Koon for the Rodney King beating, by way of justification he cited the fact that the accused policemen had "families" to care for whereas King was simply a man charged drunk driving and resisting arrest (Newton 1993). "Family" has always been an alibi if not a licence for visiting social others with unremitting violence. Such racist thinking is the culmination of the revival of scapegoating Black family households for the poverty of African Americans. It started up again with Bill Moyers's televised documentary entitled, "The Vanishing Black Family—Crisis in Black America" (broadcast January 26, 1986) and has—once again—cast African Americans beyond the pale of Americanness and acceptable humanity.[6] Then there is the wave of anti-smoking ads that inundated the airwaves of California in winter and spring of 1993. When they didn't target single or working-class women, they took aim at Latinos, as in the spot that featured a dull-eyed Latino man smoking continuously in front of the tv set oblivious to the coughing of the little girl sitting next to him. To no one's surprise, then, a comforting consensual "us" gives way to an aggressive "us" and "them." A message ostensibly about health turns out to be always already one about class, race, and gender in the form of "family": the working-class non-familial intruder (clearly the lover of the girl's mother) who threatens not just the physical but also the moral well-being of the home constitutes a warning addressed to the Latina mother who tolerates his presence and his presumably male working-class indifference to matters of health and care of children.[7] In the mouths of the defenders of the New Consensus, "family" and "health" are the kinder and gentler codewords of '80s and '90s social conservatism replacing the stern "law and order" slogan that punctuated speeches by George Wallace and Richard Nixon in the '60s and '70s.

These recent examples tip us off to how this discourse operates, namely, that handwringing and expressions of alarm over the decline or death of The Family having always been a tactic for reinscribing and protecting the so-called normative "humanity" of (straight) upper-middle-class whites through stigmatizing social others for lack of "family." As Georges Canguilhem remarked about norms and normativity,

> The normal is then at once the extension and the exhibition of the norm. It increases the rule at the same time that it points it out. It asks for everything outside, beside and against it that still escapes it. A norm draws its meaning, function and value from the fact of the existence, outside of itself, of what does not meet the requirement it serves. The norm is not a static or peaceful, but a dynamic and polemical concept. (1989: 239)

And I want to argue that while such an imperializing dynamic between self and other, between the normal and the pathological has always lain at the heart of the familial discourse, it has rarely resulted in simple and stable social geographies; on the contrary, I want to suggest that this discourse of anxiety and fear has produced social distinctions and identities only insofar as they are made to seem weak, permeable, and vulnerable to violation or colonization by the "outside." The "them" is always potentially lurking within "us." Thus it should come as no surprise that since its inception familial discourse has always constructed the humanizing "family" as under threat and in "need" of constant nurturing, surveillance, and public and private intervention.

Genealogies: Culture Wars and "Keeping Human Beings Human"

This discourse is actually a very old story (Reid 1993). It has a genealogy worth tracing, if only briefly, that stretches back to the eighteenth century when the normative family household was first elaborated and inscribed on the social body of Western European societies and their colonies. Then as now, therein lay the sense of what it meant to be "human," and it was the domestic family and its sentimental narratives that grounded and lent meaning to all other social arrangements. As Theodore Roosevelt was to put it at a much later date:

> But it is the tasks connected with the home that are the fundamental tasks of humanity. After all, we can get along for the time being with an inferior quality of success in other lines, political or business, or of any kind; because if there are failings in such matters we can

> make good in the next generation; but if the mother does not do her duty, there will either be no next generation, or a next generation that is worse than none at all. In other words, we cannot as a Nation get along at all if we haven't the right kind of home life. Such a life is not only the supreme duty, but also the supreme reward of duty. Every rightly constituted woman or man, if she or he is worth her or his salt, must feel that there is no such ample reward to be found anywhere in life as the reward of children, the reward of a happy family life. (*The Family:* iv)

As the eighteenth century drew to a close, new familial and sexual identities superseded other social and political ones; bodies began to receive different inscriptions. The question was less, "What is your station?" or "To what estate, clan, or social caste do you belong?" or "Who is your master?" but rather "What kind of father or mother are you?" "Are your children healthy and well-behaved?" "Do you love your parents?" "What are your earliest childhood memories?" Sexual practices became an obsession and other questions began to be asked: "What do you do in the secret of the night?" "Do you touch yourself?" "What is your sexual life like?" "Who do you desire?"

A culture war began to be waged in the name of the new domesticity based on heterosexual monogamy, love matches, and sentimental intimacy focused on children. In ways not dissimilar to present-day familial discourse, writers tried to mobilize citizens by uncovering enemies without and within the household. Now, in the late eighteenth century and early nineteenth century, the first group to fail the human test of "family" in Europe and North America was of course the metropolitan and colonial aristocracies and their willful middle-class imitators: their wasteful life-styles—their bodily practices—were castigated as being utterly inimical to the new "family values." Attacks on arranged marriages, the forced celibacy of apprentices, servants, and Roman Catholic clergy, libertinage, mercenary wet-nursing, and so forth discredited the noble body of alluring surfaces, power, and pleasure; and it delegitimated the social and gender relations for which that body served as relay and support. Elite women were singled out for living "selfish" lives that were too free and for preferring a life of public power and sexual autonomy to the charms of the endless daily tasks of the new child care. And writers accused men of ignoring their families and throwing away their virile energies in sexual pleasure, gambling, and lavish dinners. Philosophers, demographers, physicians, and, not least, novelists concocted narratives that told melodramatic tales of the lack of so-called true family life among the nobility and of how traditional aristocratic society (especially its women) blocked the fruition of new domestic families and renewal of the nation.

This century-long culture war culminated in substituting for the discredited noble body of surfaces the body of depth of the liberal subject. This new body enclosed legible gender difference, heterosexuality, and familial sentiment and interiority, universal human qualities presumably available to one and all, provided that one adopted the imperatives and practices of domesticity. It was that body that familial discourse at once appealed to and tried to produce through narratives of endangered domesticity. Anxiety, fears, and tears became the hallmarks of a new, emergent familial sensibility and desire, of a new subjecthood and citizenship.[8] It was through "family" and the middle-class identities it afforded that a new democratic imagined community of postfeudal and postrevolutionary Europe and United States was forged. As Sarah J. Hale, American author of the domestic manual *Manners; or, Happy Homes and Good Society* (1868) put it,

> What composes our country, and makes its life? Not its wide prairies, with their billows of undulating green; not its lofty mountains, with their and inexhaustible treasures of ore; not our vast oceans, rolling rivers, nor swelling streams. Grand as each and all of these may be, they are not the country's life, though affecting that life,—not the country's power, though increasing that power: its true life must reside in the Home; *for it is the aggregate of homes which make up the country, and it is from them that all good must flow which governs and regulates the nation.* Hence the importance of making our homes the center of happiness, usefulness, and intelligence. The higher the home standard, the higher the influence exerted by those who go forth from that home.[9] (1972: 115–16, my emphasis)

In the nineteenth century came the turn of urban workers, disenfranchised peasants, the enslaved, and the colonized: public health officials, social philanthropists, journalists, anthropologists, and politicians constructed these groups and their bodies as disorderly, biologically misfit, devoid of any sense of family life, and, eventually, as pathologically deviant. The relentless repetition of these cultural narratives defined the liberal body of the white metropolitan and colonial middle classes against everyone else, and the middle classes' embodiment of domesticity stood as the sign of their exemplary humanity, and the absence of "family" among the peasants, the urban workers, the enslaved, and the colonized designated their social and subjective existence as abject. Sarah Hale, as in everything, was perfectly clear about this. Speaking of the "original" family of Adam and Eve, she proclaimed that the United States' Manifest Destiny was familial:

> But one race retains the Eden laws of love and home; and in that race only is the faith and the worship of the true God. From that race were the families that settled and made our American people. In two centuries and a half, this North-American empire has gained power and place in the great family of nations: compared with her, those old cradles of civilization and centers of knowledge and glory—Asia and Africa—are now only blanks in the lot of humanity. (Hale 1972: 24)

However, because of these abject groups, it was understood that the new family life was a very fragile thing. Middle-class private life was rarely experienced as secure but under constant threat from without: the street could lure husband or adolescent away with promises of pleasure and ambition, crowds could well up in revolutionary fervor and swallow alive whole families and social classes; the uncertainties of the workplace and market place endangered the well-being of honest households, and biological contagion was borne along upon the bodies of the unwashed. Weak and vulnerable as it was constructed and as it is still portrayed today, the normative family and its healthy bodies nonetheless stood in the eyes of many a politician, charity worker, and public health officer as the pillar of society and the sole barrier standing between social order and anarchy. François Guizot, Louis-Philippe's prime minister who was deposed along with his king in the 1848 revolution, drew the following conclusions from the political turmoil in his treatise, *On Democracy* (1849), which he wrote while in exile:

> The family is now more than ever the first element and last bulwark of society. . . . Our great cities, their hustle and bustle of business and pleasure, and the temptations and disturbances that they spread ceaselessly, would soon throw society in its entirety into a state of ferment and deplorable laxity if domestic life everywhere present throughout the territory, its calm activity, its permanent interests, and its immutable ties did not oppose this peril with solid barriers. (1849: 138–39, my trans.)

For Guizot, as for many nineteenth-century writers, male and female, the regulative family household provides an irreplaceable refuge for the cultivation of human sentiment and interiority—weapons of last resort—within the liberal body: "Virtues abound in the family and good feelings in hearts. We have the wherewithal to fight against the evil which is devouring us" (1849: 156–57). If you substitute the 1960s for the 1848 Revolution, Guizot's discourse echoes pronouncements by Dan Quayle, George Gilder, and the Christian Right but also by liberal proponents of the New Consensus of today.[10]

Now, if there were many potential enemies standing on the borders of the domestic family, just as many if not more emerged treacherously from within: from the latter half of the nineteenth century and down to the present, psychiatrists, experts of all kinds, and the media have constructed narratives of middle-class households also "endangered" by wayward hysterical mothers and daughters, effeminate sons, unmarriable cousins, sexual "perverts," emancipated women, and maniacal or impotent fathers.

Since then the dominant cultural narratives have been of family life disrupted or destroyed by all manner of causes, internal and external. As a consequence, in everyday discourses—learned and popular—no one family household, profession, social class, gender identity, or political arrangement stands free of the dangers of pathological deviance or lapses issuing from within or descending from without. In this paranoid universe of dying family households, nothing is settled, once and for all: the frontier between the normal and the pathological remains shifting and porous enough such that threats to "family" may erupt anytime, anywhere. Good fathers may turn out to be workaholics or worse, bisexuals, admired teachers child molesters, dutiful mothers sickly or in today's updated version, simply a smoker and a drinker ("doing drugs in the womb" titled an *LA Times* editorial);[11] a favorite daughter may lead a double life as a prostitute (the nineteenth-century version) or one day be a pregnant runaway (in the late twentieth century) or a promising son may be suffering from "teenage syndrome." "Family" and the social and gender identities that turn on it are always already in jeopardy, on the verge of disappearing for good, and thus always desired as an absence.

Yet while the effect of these stories has been to destabilize the domestic family, one thing has remained constant: "family" and its liberal body have remained the very measure of the "human," that is to say the bedrock of acceptable social and individual existence according to the straight white middle class. This has also been the double role of psychoanalysis' "familialist reduction" (Deleuze and Guattari 1983: 270) with respect to bodies and the desires they were thought to enclose: as Foucault once put it, the new science of desire could issue a call to the middle classes to bring their children to the psychoanalyst's office, for the parents could rest assured that what they would learn was that at bottom what the wayward children really desired all along was Mommy-Daddy (Foucault 1980: 113).

Technologies of Subjectivity and the Social

So the cry "death of the family" is a time-worn refrain almost two centuries old. It began long before mass culture circulated through daily newspapers, tv, and physicians' waiting rooms; it has amounted to a relentless campaign to "make and keep human beings human" through inscribing lack of "family" on and beneath the skins of our bodies. But far from being the fundamental condition of desire and subjectivity (as Lacanian cultural critics would have us believe), "family" and "lack" are no more than banal but still powerful discursive mechanisms and tactics for articulating multiple bodies, desires, and social practices in specific ways around the unitary, normative pole of domesticity. In this fashion do the discourses of the decline or death of the regulative family household constitute a massive technology of subjectivity and the social.

However, the mobilization of bodies and desires for "family" has been no less powerful because of its discursive construction. Indeed, its power is discursive to the core. In familial discourse the production of "family" as desirable norm has always required freaks and outcasts that name the norm indirectly by virtue of their departure from it. This sideshow of familial others has had two powerful, advantageous effects. First, it has allowed the norm to remain invisible, thus rendering the normative family more impervious to the vagaries of individual and collective experience or to "empirical" refutation. As a consequence, catastrophic private or social events have tended not so much to discredit "family" as to renew calls for its protection and revival. Second, it maintains the norm just out of reach, something forever to be desired but never quite realized. Thus, in constant surveillance of ourselves and others—"who's next?"—we stoke our desires of "family" with tales of its disruption or decline in hope that someday we will nonetheless "get it right" in our daily lives. You might say that in this way the so-called modern domestic family has never seemed to fully work either; it has been always already in crisis, internally weak, and open to "invasion" by outside forces. But rarely have narratives of crisis been turned around in public discourse to call the normative family household in question. As Gary Indiana put it, speaking of child abuse and the anti-abortion climate,

> The connection is never made that the creation of a family is not an automatic or desirable destiny of every "normal" person, that other kinds of lives could be equally or more fulfilling, or that unwanted babies grow up to become the bogeymen of "crime" because they were victims of crime from the moment of birth. Instead, American

culture encourages the ugly wish that all its preborn fetuses will come to term, and that all the males will grow up to be Oliver North. (1990: 146–47)

Where's "Family"?

Now today, sociologists are hard put to locate the domestic family and its human bodies, anywhere. They can't find it even among the white middle classes. Something like less than 14 percent of households in the United States fulfill the old domestic norm of male breadwinner, female homemaker, and children; the figure drops to 5 percent if the additional factor of never having been married before is thrown in (Anderson: 1987). So where's the family if not in the household? Nancy Armstrong, a feminist critic, has claimed that "the most powerful household is the one we carry around in our heads" (1987: 251). Yet even the desiring bodies of its members seem to be less and less sites of family; these days, the drift of desire is such that household members are only *playing* at Oedipus, convincing fewer and fewer people. The Oedipal narrative of Mommy-Daddy-Me, or better still of the Phallus, isn't faring well lately (Deleuze and Guattari 1983: 356). This secret "truth" of the human subject is decidedly a thing of straight males; Deleuze and Guattari write:

> The news travels fast that the secret of men is nothing, in truth nothing at all. Oedipus, the phallus, castration, "the splinter in the flesh"—that was the secret? It is enough to make women, children, lunatics, and molecules laugh. The more the secret is made into a structuring, organizing form, the thinner and more ubiquitous it becomes, the more its content becomes molecular, at the same time as its form dissolves. (1987: 289)

Well, if members' inner desires don't embody the human and the familial so well anymore then perhaps we'll have to look elsewhere: one favorite place over the years on tv and in film (since Lassie) have been household pets: looking for family feeling or want to know the barometer of family health? Cut to pooch-Oedipus there on the couch that smiles broadly or whimpers as needs be. But we can take it that this is not a good sign when animals, as domestic as they may be, are responsible for keeping human beings human through family.

Still, as the culture wars here and elsewhere testify, the old domesticity has a long life. In this regard two British feminists, Michèle Barrett and Mary McIntosh, have suggested that domesticity and its norms are less likely to be encountered in the home than in the public sphere: they remind us that even as daily practices depart from the regulative family,

"family values" seem ever more omnipresent: from the division of labor and job segregation in the workplace, school curricula, insurance policies, health care delivery, welfare regulations, advertisements, and travel brochures to films, tv soap operas, newscasts, mail-order catalogs, etc. (1982: 28–34). But then again, perhaps that is where these values have always been all along and less so in households. Certainly many of the examples I've presented in this essay would seem to confirm this. Historian Stephanie Coontz suggests that daily practices in this country never have embodied the normative family.[12] In any case, this explains in part the hysteria over lifting the ban on homosexuals in the military; on some level most realize, if only dimly, that domesticity can't be counted on to guarantee heterosexual identity. The role devolves more and more to other public and private institutions.

"Family" has never been a private matter. The normative family household and the bodies of its members have never constituted either a sanctuary of values and relations at a safe remove from the outside or a depth of feeling and sentiment in opposition to the public sphere and the bodies of social others. Rather, they have always been a projection from a surface of inscription (paper, bodies, spaces) upon which discourses of "family" and liberal subjectivity have attempted to write themselves. "Family," like gender, doesn't simply exist; it must be a non-stop public performance. Only now, The Family has been turned inside out: to find "family" these days you had better turn on the tv, run to court, go to school, attend a "Family of Pride" gay and lesbian parade in Southern California, or maybe, like the young parricides of Rush Springs, Oklahoma, stop by McDonald's. Or better still, read safer-sex pamphlets distributed by the Surgeon General's office, mainstream health clinics, and hospitals; there, the process of inscribing bodies and spaces with "family" proceeds apace with murderous force; they promote the notion that "safe-sex" [*sic*] actually exists but only among monogamous, heterosexual couples (Patton 1990: 99–104). Better dead than "deviant," I guess. In this literature collective murder-suicide is in the offing. It might just happen. The depths of regulative households and straight, liberal subjects may disclose a tomb, not only for others (an old story of "them") but also for "us," the ever-mythical "general population."

In the Meantime

In the present state of affairs, then, it should come as no surprise that the Dan Quayles are at it again, be they neo-cons, neo-liberals, or even progressives, editors of *Commentary,* the *New York Times* or *Dissent.* These are nasty times. However, as Halberstam and Livingston have suggested in the introduction to this volume, some of them may have lost their touch: for example, rather than simply denouncing others for not embodying the human qualities of standard domesticity, the Republicans in the 1992 convention made the mistake of also trying to define "family," that is, they made *their* version of "family" visible and by so doing forced viewers to compare it to their own situation. Viewers may have decided that it's simply not for them. The "death of the family" may no longer tug at the heartstrings anymore or cause people to sweat in the old way while watching tv at home or flipping through magazines in the doctor's office. Perhaps many are finding it boring. Or—though some people may find it too glib to say—maybe people are *on* to the old discourse: it's just a mechanism for scaring people during hard times and for telling them what to do and who to be. Perhaps more and more of them are staking their identities elsewhere in other discourses and bodily practices.

However you look at it, what is going on now is no simple matter. You can argue that it is virtually impossible to evoke household and sexual practices without some reference to conventional norms. But if normative discourse has begun to implode, if we are reaching some postfamilial, posthuman moment, what is there to say? Deleuze and Guattari make a suggestive answer when they claim that "there are only inhumanities, humans are made exclusively of inhumanities, but very different ones, of very different natures and speeds" (1987: 190). If the "non-human" dwells in the house of humanity, what next? What happens when the sideshow of familial queers is the only show in town? When zombies are the majority among the "living"? What might be at hand is not so much a self-contradiction that gives "birth" to a new era, much less a self-deconstruction as a lateral shift into a different space, that of "diversity," with different dynamics, affects, and intensities, verbal and bodily. "Lack" of "family" will be missed only mildly except by the old gatekeepers of the human whose power and arsenal of cultural and institutional weapons remain frighteningly real.

Some bodies, some households. Bodies without Law, households without norm. Postmodern families would not be families at all. Or only partially; but who, when? Some members but not others? Some part of the day but not the full day? When one eats, sleeps, or fornicates? When one works (for the household?)? And what of the body? When investment drifts away from the master organs, you know it can go almost anywhere.

New bodies are at large—and at "home." They are less and less "prisoners of the soul," of those sexualized childhood memories that are the "truth" of our selves to social experts, teachers, and politicians and to which they may appeal in times of social crisis. Those inner well-springs of social discipline, inscribed in the depths of the liberal body, are drying up. That's what Christopher Lasch was surely referring to by his lament of the atrophy of affect (1978). He wasn't talking of affect as such or of intensities but rather of *familial sentiment* and its mechanisms of nostalgia and guilt.[13] Those bodies still gendered "female" (lesbian, straight, or queer) ask less and less of men in power and seek even less to please them; and they rarely look up to those women who employ, supervise, and manage. Cross-dressing and gender-bending are the newest games in town. And tattoos are threatening to edge out makeup and clothes as self-inscriptions of choice. These bodily practices are to please yourself and are lateral pleasures for others who can collect them as they wish but, here, voyeurism is no longer guilty theft. Of the "men" straight males are always slowest to change but even then some are trying a casual indifference to the dangers and excitement of patriarchal abuse and recruitment. They don't expect and no longer desire the same old initiation and tutelage. Blood debts are losing their power to fascinate.

Once they get out, these bodies aren't interested in going "home" very much either, for family households aren't havens in a heartless world as some claim (Lasch 1977). Reflecting upon the fact that most murders are committed by friends and family members, Sue Grafton's detective Kinsey Millhone mused, "a chilling thought when you sit down to dinner with a family of five. All those potential killers passing their plates" (1987: 6).[14] It's safer in the streets among strangers where queer zombies roam.

Notes

I have benefited from continuing conversations with Linda Brodkey and Lisa Bloom whose interest in new forms of writing has crucially shaped successive drafts of this essay.

1. For the report's call for the creation of a "Family Strength Index" by the Census Bureau see *Families First:* 62–63.
2. The Menendez brothers case tried in Los Angeles presented the inverted image of this one: Lyle and Erik Menendez shot their wealthy parents with shotguns over 10 times. The prosecution accused them of killing in order to consume whereas they claimed that they had been raped by their father and killed out of fear for their lives. The defense seems to have worked, for the trial ended with two hung juries. See Abrahamson 1993: A1.
3. On "toxic parents" see Forward 1989.
4. This repeated stress on the two-parent household leaves its members in a position of extreme vulnerability; from this perspective, according to Stephanie Coontz, it doesn't take much to send individuals, especially children, and entire households spiraling to their ruin: "It means that any child is only one

death, one divorce, one blood test away from having nothing" (Coontz 1992: 230).

5. Even presumably liberal publications like the *Atlantic* have lent ample space to writers who defend Quayle's position in the name of the New Consensus; see Whitehead 1993; and most recently Health and Human Services Secretary Donna Shalala, one of the most liberal members of the Clinton Administration, when asked whether Murphy Brown was right in bearing a child, replied, "I don't think anyone in public life today ought to condone children born out of wedlock . . . even if the family is financially stable." See Eaton 1994 and the approving *Los Angeles Times* editorial "The Instant Formula for Poverty."
6. The first wave of scapegoating actually began in 1965 with the publication of the Labor Department study, "The Negro Family: The Case for National Action," directed by the then liberal Harvard sociologist Daniel Patrick Moynihan. On Moynihan's report and scapegoating Black family households see "Scapegoating" and Coontz 1992, ch. 10. More recently, *Newsweek* proclaimed that a new consensus among scholars now vindicates Moynihan's original thesis; it was featured as a lengthy cover story, "World without Fathers: The Struggle to Save the Black Family" that contained two articles, "Endangered Family" (pp. 16–27) and "Protecting the Children" (pp. 28–29) accompanied by no less than *eight* human interest stories; see "World without Fathers." Finally, Moynihan's revived thesis is now embodied in the 1994 Clinton welfare reform bill that punishes recipient mothers for having more children (even though their households average fewer children than the national mean); when the Maryland NAACP voted to study Clinton's bill, it was interpreted by the triumphant *Los Angeles Times* as a long-overdue capitulation of African-American leaders to Moynihan's "facts"; see the editorial "Moynihan the Prophet."
7. I want to thank Linda Brodkey and George Mariscal for their acute insights into reading this ad.
8. For the United States consult D'Emilio and Freedman 1988 and Coontz 1988; for Great Britain see especially Weeks 1981 and Armstrong 1987; for France see Reid 1993.
9. I want to thank Nicole Tonkovich for bringing this book to my attention.
10. For a helpful synopsis of familial thinking by the New Right in the U.S. and U.K. consult Abbott and Wallace 1993.
11. See "Doing Drugs."
12. See Coontz 1992. Both of Coontz's books (Coontz: 1988, 1992) are exemplary in their corrosive empiricism; they dispel once and for all any simple, linear notion of family household history for the United States and the idea that any one family household was embodied across all communities and social classes. Normalizing narratives of private life simply cannot account for the sheer complexity and diversity of household arrangements from colonial times down to our day in different classes, regions, and communities. The Ozzie and Harriet ideal was a transient moment at best in post–World War II

middle-class suburbia. In my view what her account omits is how *discourses* have nonetheless constructed a dominant norm that has continuously articulated bodies, social relations, and subjectivities in terms of middle-class domesticity for almost 200 years. In short, she overlooks the distinct possibility that "family" has always existed normatively, and even especially so, while daily practices were at variance with its strictures and imperatives.

13. However, the "self-esteem" movement may offer new hope to welfare activists, social experts, and government officials despairing of being able to discipline citizens; see Cruikshank 1993.

14. Some recent statistics on rape offer no reassurances either. A revisionist study has claimed that only 22 percent of rapes are committed by strangers; 29 percent by non-relatives known to the victims; 9 percent by a boyfriend or former boyfriend; 16 percent by a relative not of the immediate family; 11 percent by a father or stepfather; 9 percent by a husband or a former husband; see "Survey Shows."

Works Cited

Abbott, Pamela, and Claire Wallace. 1992. *The Family and the New Right.* London: Pluto Press.

Abrahamson, Alan. 1993. "Lyle Menendez Admits Lies, Insists He Killed in Fear." *Los Angeles Times,* 22 Sept. 1993: A1.

Anderson, Lauren. 1987. "Property of Same-Sex Couples: Toward a New Definition of Family." *Journal of Family Law* 26.2 (Mar. 1987): 357–72.

Armstrong, Nancy. 1987. *Desire and Domestic Fiction: A Political History of the Novel.* New York: Oxford University Press.

Barrett, Michèle, and Mary McIntosh. 1982. *The Anti-Social Family.* London: Verso.

Canguilhem, Georges. 1989. *The Normal and the Pathological.* Trans. Carolyn R. Fawcett with Robert S. Cohen. New York: Zone.

Carter, Bill. 1992. "Networks Unite on Child Abuse." *New York Times,* 26 Aug. 1992, late ed.: C13.

"Children of the Cult." *Newsweek,* 17 May 1993: 48–50.

Clinton, President Bill. 1993. Presidential News Conference of April 20, 1993. *Weekly Compilation of Presidential Documents* 29.16 (April 26, 1993): 625.

Coleman, Brenda C. 1993. "Study Says 1 in 50 Kids May Have Jailed Parent." *San Francisco Examiner,* 11 Aug. 1993: A10.

Coontz, Stephanie. 1988. *The Social Origins of Private Life.* London: Verso.

———. 1992. *The Way We Never Were: American Families and the Nostalgia Trap.* New York: Basic Books.

Corwin, Miles. 1993. "To Catch a Killer." *Los Angeles Times,* 11 Aug. 1993: B1.

Cowley, Geoffrey. 1993. "Who's Looking after the Interest of Children?" *Newsweek,* 16 Aug. 1993: 54.

Cruikshank, Barbara. 1993. "Revolutions Within: Self-Government and Self-Esteem." *Economy and Society* 22.3 (Aug. 1993): 327–44.
"Dad Kills Wife, Two Kids." *New Haven Register,* 12 July 1993: 8.
"Death Wish: The Day of Judgment." *Newsweek,* 3 May 1993: 22–31.
Deleuze, Gilles, and Felix Guattari. 1983. *Anti-Oedipus.* Vol. 1 of *Capitalism and Schizophrenia.* Trans. Robert Hurley, Mark Seem, and Helen R. Lane. Minneapolis: University of Minnesota Press.
———. 1987. *A Thousand Plateaus.* Vol. 2 of *Capitalism and Schizophrenia.* Trans. Brian Massumi. Minneapolis: University of Minnesota Press.
D'Emilio, John, and Estelle B. Freedman. 1988. *Intimate Matters: A History of Sexuality in America.* New York: Harper and Row.
"Doing Drugs in the Womb." Editorial. *Los Angeles Times,* 18 Sept. 1993: B7.
Eaton, William J. 1994. "Shalala Revives 'Murphy Brown' Pregnancy Issue." *Los Angeles Times,* 15 July 1994: A1, A22.
Families First. 1993. Report of the National Commission on America's Urban Families. Washington, D.C.
The Family: Preserving America's Future. 1987. A Report to the President from the White House Working Group on the Family. Washington, D.C.
Forward, Dr. Susan, with Craig Buck. 1989. *Toxic Parents: Overcoming Their Hurtful Legacy and Reclaiming Your Life.* New York: Bantam.
Foucault, Michel. 1980. *An Introduction.* Vol. 1 of *The History of Sexuality.* New York: Vintage.
Grafton, Sue. 1987. *A Is for Alibi.* New York: Bantam.
Guizot, François. 1849. *De la démocratie.* Paris: Victor Masson.
Hale, Sarah J. 1972. *Manners; or, Happy Homes and Good Society All the Year Round.* 1868. Rpt. New York: Arno Press.
Indiana, Gary. 1990. "Blood Brothers." In *Democracy: A Project by Group Material.* Ed. Brian Wallis. Seattle: Bay Press, 141–47.
"The Instant Formula for Poverty." Editorial. *Los Angeles Times,* 16 July 1994: B7.
Lasch, Christopher. 1977. *Haven in a Heartless World: The Family Besieged.* New York: Basic Books.
———. 1978. *The Culture of Narcissism: Life in the Age of Diminishing Expectations.* New York: Norton.
"Leaving Two Home Alone, Parents Are Held." *New York Times,* 30 Dec. 1992, late ed.: A10.
Mills, David. 1991. "Oprah, Children's Crusader." *Washington Post,* 13 Nov. 1991: A1.
"Mother Goes to Jail for Baby Food Tampering." *FDA Consumer* 26.7 (Sept. 1992): 44.
"Moynihan the Prophet." *Los Angeles Times,* 14 March 1994: B6.
Nash, Nathaniel C. 1993. "Argentines Free 268 Children Reportedly Held by Sex Cult." *New York Times,* 3 Sept. 1993, natl. ed.: A1.
Newton, Jim. 1993. "Koon, Powell Get 2 1/2 Years in Prison." *Los Angeles Times,* 5 Aug. 1993: A1.
Patton, Cindy. 1990. *Inventing Aids.* New York: Routledge.

"Pleas for Protection of Clinics Follow Shooting of Abortionist." *San Diego Tribune,* 21 Aug. 1993: A11.

Popenoe, David. 1993. "Scholars Should Worry about the Disintegration of the American Family." *Chronicle of Higher Education,* 14 April 1993: A48.

"Quayle Deplores Eroding Values; Cites TV Show." *Los Angeles Times,* 20 May 1992, San Diego County ed.: A1.

Reid, Roddey. 1993. *Families in Jeopardy: Regulating the Social Body in France, 1750–1910.* Stanford: Stanford University Press.

"Scapegoating the Black Family." *Nation,* 24–31 July 1989 (Special Issue).

"*Shattered Trust: The Shari Karney Story.*" NBC, 27 Sept. 1993.

Smothers, Ronald. 1992a. "Big Molestation Trial Nears Its Close." *New York Times,* 23 March 1992, late ed.: A18.

———. 1992b. "Day-Care Owner Is Convicted of Child Molesting." *New York Times,* 23 April 1992, late ed.: A16.

———. 1993a. "In a Day-Care Case, New Questions, Few Answers." *New York Times,* 23 July 1993, natl. ed.: B9.

"Survey Shows Number of Rapes Far Higher than Official Figures." *New York Times,* 24 March 1992, late ed.: A14.

Timnick, Lois. 1990. "Buckey Jury Deadlocks; Mistrial Is Declared." *Los Angeles Times,* 28 July 1990: A1.

Verhovek, Sam Howe. 1993. "Town Rallies Behind Boys Who Killed Their Father." *New York Times,* 25 July 1993, natl. ed.: 1–9.

Weeks, Jeffrey. 1981. *Sex, Politics, and Society: The Regulation of Sexuality since 1800.* London: Longmans.

Whitehead, Barbara Dafoe. 1993. "Dan Quayle Was Right." *Atlantic* 271.4 (April 1993): 47–84.

Williams, Lance. 1993. "Racist Admits '87 Slaying." *San Francisco Examiner,* 11 Aug. 1993: A1.

"World without Fathers: The Struggle to Save the Black Family." *Newsweek,* 30 Aug. 1993: 16–29.

PART IV:

TERMINAL BODIES

TEN

Reading Like an Alien: Posthuman Identity in Ridley Scott's *Alien* and David Cronenberg's *Rabid*

Kelly Hurley

The general concern of this chapter is what *Screen* identified, in a 1986 special issue title, as the cinema of "Body Horror": a hybrid genre that recombines the narrative and cinematic conventions of the science fiction, horror, and suspense film in order to stage a spectacle of the human body defamiliarized, rendered other. Body horror seeks to inspire revulsion—and in its own way, pleasure—through representations of quasi-human figures whose effect/affect is produced by their abjection, their ambiguation, their impossible embodiment of multiple, incompatible forms. Such posthuman embodiments are liminal entities, occupying both terms (or rather, existing in the slash between them) of the opposition human/not-human. I will focus on posthuman embodiment in David Cronenberg's *Rabid* (1977) and Ridley Scott's *Alien* (1979), reading these as texts that work to confound and exceed the biological, sexual, and psychic categories through which we traditionally construct a "human" identity.

"Body Horror" as described above is nothing new in cinema; representations of not-quite-human embodiments can be abundantly found in mid-century horror/SF, in such films as *The Fly* (1958), *Creature from the Black Lagoon* (1954), *The Alligator People* (1959), *The Creeping Unknown* (1956), *The Slime People* (1962), and countless others. However, there seems to be some critical consensus that body horror has turned itself up several notches in recent decades,[1] manifesting a more lurid sensationalism since the 1970s, the time of *Rabid* and *Alien*. One sees a tendency toward ever more disgusting content (the cannibalism films of George A.

Romero, Wes Craven, and Tobe Hooper), ever more sadistic violence (the *Friday the 13th* and *Halloween* series), ever more graphic, painstakingly rendered corporeal transformations (*The Howling, An American Werewolf in London,* both 1981), ever more grotesque and unspeakable embodiments (John Carpenter's 1982 remake of *The Thing,* Cronenberg's 1986 remake of *The Fly*).

As my choice of adjectives in the last sentence is meant to illustrate, recent body horror, though it has met with a respectable share of critical attention, has not always met with approval. Pete Boss, for example, argues that the "bodily destruction of the modern horror film," compared with the more restrained horror of the past, is "often casual to the point of randomness; devoid of metaphysical import, it is frequently squalid . . . [and] mechanically routine in its execution . . . " (16). Modern body horror, by this account, is meaningless, excessive, gratuitous; and as such it may be harmful to the consumer as well. Harvey R. Greenberg claims that "horror beyond psychological tolerance cancels enjoyment. . . . In cruel cinema, any possibility for a healing catharsis is deliberately sacrificed in favor of overwhelming the viewer's capacity to endure psychic pain" (84, 85).

The sheerly horrific quality of body horror, however, has been justified within an ideological framework. Films like *The Texas Chainsaw Massacre* (1974), *The Hills Have Eyes* (1977), and Romero's zombie trilogy have been praised as scathing critiques of American consumerism and overinvestment in the nuclear family.[2] Alternatively, even when body horror fails to deliver such a critique—or contradicts and undermines its own attempts to forward a "progressive" agenda—its very failure can be said to reveal the mechanisms of resistance that hold normative cultural institutions in place. *Alien* has been the object of an extraordinary amount of analysis along these lines, as a text that initiates but cannot follow through on a Marxist and/or feminist critique of American culture. Within these terms, the horror film serves as a document of the cultural unconscious. The text must be psychoanalyzed: its apparent content must be stripped away, its repressed content must be uncovered and interrogated, in order that we may understand the *real* fears that horror mediates for producer and consumer.

I should state here that (for the space of this chapter at least) I am not interested in the politics of body horror, progressive or otherwise. In fact, it is my contention that the horror film has been asked the question "What are you repressing?" too often, to the exclusion of other questions; that the reading attending too scrupulously to the displaced ideological content of the horror film, like the reading that dismisses it as "squalid," fails to take into account what the horror film *can* accomplish that other genres cannot.[3]

The narrative told by body horror again and again is of a human subject dismantled and demolished: a human body whose integrity is violated, a human identity whose boundaries are breached from all sides. One needs to place this narrative, and the images from which it is generated, within the discourse (and culture) of postmodernity, particularly as that discourse theorizes the breakdown of human specificity and the erosion of human identity, embodied and otherwise. Here I am concerned less with the reputed postmodern fragmentation of human identity than with its reconfiguration through the pluralization and confusion of bodily forms. In horror films, with their standard figures of the human/not-human (alien, mutant, cyborg, psychopath), the trope of bodily ambiguation has hardened virtually into a filmic convention. At the same time, the sophisticated special effects technologies that have recently transfigured the genre—and indeed, the very outrageousness of a genre whose "legitimate" goals are shock and disgust—means that the horror film is able to effect a spectacular visual staging of bodily ambiguation that can only be entertained conceptually in theoretical discourses on postmodernity. I wish to take the horror film literally for the moment, focusing on monstrosity as spectacle rather than metaphor, and on body horror as a speculative narrative that sets out new economies of identification and desire, rather than leading us back, in allegorical fashion, to the ones we already know too well. I am not interested in how horror helps us manage our fears, but in how it helps us to imagine otherwise, outside of the parameters of "the human," in its generation of posthuman embodiments both horrific and sublime.[4]

Psychoideological Readings of Horror

A fairly standard line on horror is that the genre enables its consumers to confront, temporarily and vicariously, their most basic fears and/or unconscious desires. For example, we enjoy watching aliens and axe-murderers rend human victims from limb to limb, or watching human subjects undergo prolonged and obscene bodily metamorphoses, because this enables us to acknowledge our own dread of the death, disease, mutilation, and other transformations that may overtake us or our loved ones. Alternatively, if we identify with the monster and not the victim, viewing the film may serve to vent violent impulses that could otherwise erupt elsewhere, and less safely. The incursion of monstrosity in the horror text represents a "return of the repressed," of fears and desires too intolerable to be examined directly. Like its close relative the nightmare, the horror text clothes the repressed contents of the unconscious in phantasmatic narrative and spectacle, such that they may be both recognized and disavowed. Horror films, in other words, serve as cheap therapy. One enjoys the incursion of monstrosity/the unconscious from the safe distance of one's own seat; and particularly when the film effects definitive closure (the monster killed and normality restored), one goes home—or turns off the VCR—refreshed, inoculated, for the moment, against the pesky unconscious.

For the moment, I will take note of three basic assumptions that enable this model: first, the chief function of the genre is to initiate catharsis for the spectator; second, the spectator's primary relation to the text involves "identification" (and its concomitant, disavowal); and third, the spectator as conceived is fully invested in a discrete and stable human identity. These assumptions also underlie those arguments that deploy psychoanalytic methodologies in order to perform an ideological reading of horror. This move takes us from the realm of "general human" fears and desires to that of socially, historically specific ones. Within such a reading, cultural repressions are roughly simultaneous with the repressions necessitated for individual subjects, so that the types, frequency, and degrees of outright monstrosity occurring in horror film serve as indices to the contents of what might be called the cultural unconscious.

A notable and influential exposition of this argument can be found in Robin Wood's Marxian-Freudian analysis of the horror film that serves as introduction to the 1979 collection *American Nightmare.*[5] Wood, working from the distinction elaborated by Gad Horowitz, sets to one side the phenomenon of "basic repression" (the process of oedipalization that transforms the infantile body, governed only by drives, into a proto-subject ready to be inserted into the social order), here dismissed as "universal, necessary and inescapable." He focuses instead on the "surplus repression" required to transform this proto-subject into an obedient and functional member of its culture. Under capitalism, this means the disciplining of non-routinized sexual and creative energies that might prevent the subject from becoming the ideal capitalist subject, a sexually monotonous and automaton-like worker (Robin Wood 7–9). Social disciplining (the agent of which is not specified here) ensures that such energies are rerouted to the unconscious.

Of course, they do not lie quiescent there. The repressed reemerges, among other things, through the mechanism of displacement: "What is repressed (but never destroyed) in the self [is] projected outward in order to be hated and disowned" (9). The vehicle for this mechanism will be those individuals or groups discredited by or perceived as inimical to the dominant culture; within Western capitalism, these groups include women, the proletariat, non-whites, and non-heterosexuals (9–10). In other words, the contents of the individual unconscious will be discharged along the channels laid out by a normativizing culture, which disallows certain desires and practices, but admits them as the property of certain disruptive "Others," whom the normative subject is given license to despise.

Wood thus sets the stage for a cultural poetics of the horror film: "central to [the genre] is the actual dramatization of the dual concept the repressed/the other, in the figure of the monster. One might say that the true subject of the horror genre is the struggle for recognition of all that our civilization represses and oppresses . . . " (10). That is, the filmic monster serves as an expression of those desires that the spectator has been trained, through social conditioning, to disavow, and since the monster appears in the guise of "other," the spectator's disavowal may be prolonged at the same time that s/he enjoys vicariously the enactment of his/her repressed desire.

Opinions as to the political ramifications of horror-as-catharsis vary. By some accounts, horror itself, considered formally, is a profoundly conservative genre.[6] It deploys the same plot structure again and again, albeit filled with variable content. Ideological normativism is disrupted by the incursion of monstrosity and all that it represents, but the disruption is only temporary, with the horror text being at some pains to expel monstrosity and effect closure, reconfiguring the boundaries of the normative. When the monster is killed, in other words, the unconscious content it represents is sent packing back to the unconscious again. The spectator/culture enjoys the discharge of unconscious content, but does so with the understanding that this moment of catharsis will be rigidly contained and temporary. The culture is able to entertain deviations from the norm only because this happens at the level of fantasy, within the emphatically "not-real," because the moment of discharge is facilitated by a discredited, "low" genre, and because the audience is assured that the norm will be reconstituted by the end of the film. Within this account, viewing the horror film makes us bad people, complacent subjects reacclimated to cultural norms.

Robin Wood, on the other hand, while acknowledging the potential conservatism of horror (23), sees it as a genre with tremendous liberatory, even revolutionary potential. Since the horror film, uniquely, serves as a window onto the cultural and thus individual unconscious, it allows the spectator to recognize the nature and extent of his/her repressed, "monstrous" desires, acknowledge the futility and danger of projecting these desires onto social marginals, and renounce the ideological normativization that mandated both repression and projection. The "most distinguished American horror films," even in their "despair and negativity . . . are progressive in so far as their negativity is not recuperable into the dominant ideology, but constitutes (on the contrary) the recognition of that ideology's disintegration, its untenability, as all it has repressed explodes and blows it apart" (Robin Wood 23). Indeed, the very illegitimacy of the genre—since the horror film is often dismissed as frivolous—means that it can perform this deprogramming all the more successfully, for when the spectator views horror, the "censor . . . is lulled into sleep and relaxes vigilance" (13).

By this reading, the horror film has a moral duty to perform, and will be judged responsible or irresponsible on the basis of that performance. Viewing the horror film makes us better people, but only when the horror film has good intentions and carries them out scrupulously. Many later critics seems to have inherited from Wood the expectation that the horror film be "progressive" in its politics, leveling a more-or-less explicit attack on such institutions as patriarchy, capitalism, racism, and sexual conservatism. That the horror film frequently accomplishes precisely this, in startling and provocative ways, is beside the point at the moment. In any case, it avails little; being merely popular (or worse, commercial), the horror film is usually found wanting when it manifests an interest in social issues.

Alien in particular has attracted critical attention as what Greenberg calls a "sullied polemic" (101): a film that merely "pretends" to have a progressive agenda, whose seemingly radical critique of normative social roles and institutions masks an intractable underlying conservatism. For example: *Alien* seems to purvey a feminist politics by breaking gender stereotyping and offering Ripley as a character who's tougher, smarter, and more resourceful than all the men in the film. However, this surface veneer of feminism is belied by the film's deeper and perhaps unconscious level of misogyny. The Alien's[7] is a gothic version of the female body, and the film in general manifests, in displaced fashion, the horror and dread of the female genitalia that Freud described so concisely in his essay on the uncanny.[8] Alternatively, *Alien* seems to level a scathing critique of capitalism, depicting alienated workers savaged by both the pitiless Company which regards them as expendable, and the Company's double, the pitiless Alien, which regards them as prey. However, the film undercuts its own Marxist critique with a recuperative humanism: individualist as opposed to collective effort, heroic as opposed to political action, is the solution to corporate/Alien predations.[9] Or, more subtly, the film undercuts its own Marxist critique with a recuperative feminist narrative of triumphant individualism, the audience cheering as Ripley, the modern woman, defeats the Alien with technological savvy, weaponry, and pluck.[10]

My purpose here is not to discount psychoideological approaches to horror, which have produced any number of extraordinarily interesting analyses of films. My concern is rather that such approaches have delimited critical expectations of the horror film. It is revealed as a genre impoverished in its effects: endlessly and monotonously enacting narratives of repression and displacement to an unsophisticated audience compulsively returning to the theater for yet another healing (or normativizing) catharsis. More importantly, the psychoideological approach necessitates reading the images generated by body horror metaphorically (the Alien equals alienation under capitalism; the Alien equals the monstrous feminine), so that the narrative as a whole becomes a kind of allegory, with each scene pointing toward its parallel moment in the narrative of multinational capitalism, or of the traumas of oedipalization. We are led relentlessly back into the logic of "the human"—the dehumanization of life under capitalism, the drama of human sexual difference—despite the fact that the very conventions of the genre allow it to dismantle "human identity" as a construct, and generate new imaginings (however terrifying) of the human/not-human, of the human-becoming-other. The result is a reading that takes the horror film too seriously, disallowing the speculative playfulness of the genre, and, for that reason, does not take it seriously enough.

This Sex Which Is Not One

Much critical attention has been given to the sexing of *Alien*'s Alien.

> Through grotesquely emphasized erectile images, the alien insistently registers psychosexually as a threatening phallus: it unfolds itself from a seemingly inert mass into a towering, top-heavy menace; extends its insidiously telescoping jaws; slithers its tail up the leg of its fear-paralysed female victim. . . . (James H. Kavanaugh 76)

> In . . . *Alien,* we are given a representation of the female genitals and the womb as uncanny—horrific objects of dread and fascination. . . . The creature is the mother's phallus, attributed to the maternal figure by a phallocentric ideology terrified at the thought that women might desire to have the phallus. (Barbara Creed 62, 68)

> The alien's phallic identity is strongly marked (the long reptilian neck), but so is its large, expandable mouth, armed with tiers of sharp metallic teeth. As a composite image of archetypal sexual

> dreads it could scarcely be bettered: the monstrous phallus combined with *vagina dentata.* (Robin Wood 27)

Each of these sexual characterizations of the Alien, each amply and convincingly supported with evidence from the text, leads us in a different interpretive direction. The Alien's resemblance to a penis points toward a dread of masculine aggression and penetrative violence. The Alien's resemblance to a vagina or womb, and to such imaginary constructs as the *vagina dentata* or maternal phallus, points toward a dread of female sexuality, conceived as aggressive and thus castrating. The Alien's resemblance to both male and female reproductive organs points toward a general disgust with sexuality.

The issue of the Alien's "sexual identity" is further confused by its morphic variability across multiple life-cycles. A long flexible (phallic) tube emerges from a puckered (vaginal) orifice of the larval Alien; this tube is thrust down the throat of its victim in order to implant an embryo. (Kane, a man, is the only human being "deep-throated" and "impregnated" in this first film.) The developed embryo bursts from the chest of its host, looking more or less like a pulpy penis with a long tail and silver-tipped teeth. The film rarely allows the spectator more than a glimpse of the "adult" Alien, bipedal and roughly humanoid in form. Its swollen bullet-shaped head and snaky tail might seem to code it masculine, if it weren't for its mouth. Or rather, mouths: within the moist and gaping outer mouth and its ring of teeth (*vagina dentata*) lies a second mouth, a sort of oblong phallus, also tipped with teeth, that does the work of penetration. Overall, one could argue that the Alien's association with the brute physicality of coition, birth, and death places it within the register of femininity as this is said to signify the gross corporeality of the body[11]—but alternately, that the Alien's similarity to slasher/rapists of other horror films, or its doubling relation (within a Marxist reading of the film) to the Company, places it within the register of masculinist aggression and/or patriarchal capitalism.

Is the Alien a boy or a girl? Since textual evidence supports both cases, I would tend to agree with those critics who characterize the Alien as, genitally speaking, simultaneously male and female. However, I would give this argument a perverse twist and assert that the Alien is therefore neither. It is not, of course, that the question of sexual identity is irrelevant to the film, but that the Alien as embodiment (I should say, each embodiment of the Alien) exceeds the logic of a human (sexual) identity predicated on genital difference. The film, in turn, exceeds and resists an interpretation that would contain it within the field of (human) sexuality. *Alien* as narrative moves smoothly within the logic of psychoanalysis—

with its heartless computer named "Mother," and its heartless Alien bristling all over with stylized genitalia, registering all the usual signs of repression and displacement—so smoothly and indiscriminately as to overload and rupture that logic. Collapsing this foundational binarism (penis/no penis, male/female) indispensable to the constitution of human identity, the film works to take us outside of the logic of "the human," to imagine other (alien) systems of reproduction, other (alien) logics of identity. Or perhaps, it works to push us toward another (alien) logic of "the human," one predicated only occasionally and incidentally on categories of sexual difference.

Before continuing with *Alien,* I would like to consider a low-budget film that has received much less critical attention, Cronenberg's 1977 *Rabid.* Unlike *Alien, Rabid* operates fully within a semiotics of human sexuality and desire, but like *Alien, Rabid* works to overload and exhaust formations of sexual difference and identity, to represent an economy of desire and consummation that utilizes, but confounds, the usual constructions of human sexuality. These include the usual "perverse" constructions like the phallic woman, the legitimate property of horror.

Rose, the protagonist of *Rabid,* is played by porn star Marilyn Chambers (of *Behind the Green Door* fame). Rose's upper body is badly injured when the motorcycle her boyfriend is driving slams into a stalled recreational vehicle and catches fire. She is taken to a nearby plastic surgery clinic for emergency treatment, where Keloid, a partner-owner of the clinic, performs a radically experimental operation to replace her damaged skin and internal organs. As an offshoot of the surgery, however, Rose undergoes two mutations: an inability to subsist on anything but human blood, and the growth of a penis-like appendage underneath her armpit by means of which she draws that blood. Rose's victims in turn develop a disease resembling rabies. Equally ravenous for blood and slavering at the mouth, they hurl themselves on the most proximate victim, their bite carrying contagion in its turn. Eventually the imposition of martial law is able to check the disease's rapid progress through Montreal, and Rose is killed by one of her own victims. The film ends with a doberman chewing at Rose's body outside the apartment building where she died, and the credits roll as the sanitation police, collecting corpses, fling hers into a garbage truck.

We seem to have a most familiar horror narrative in *Rabid:* the indiscriminate enactment of libidinal urges threatens public order, and its agent is the aggressive, inappropriately sexual (phallic) woman. However, in the end the Law is able ferociously to recontain the unchecked desire, particularly female desire, that menaces civilization, signaling that we have returned to the regime of repression. By this reading, the film exposes (or manifests, if one doesn't like Cronenberg) the cultural repression of non-normative sexuality, which "returns" in the phantasmatic image of Rose's mutated body and the phantasmatic iterated narrative of wild-eyed, drooling people biting other people. Within psychological/horrific conventions of the displacement of repressed content, desire equals disease, and desiring women, in a double displacement that bespeaks an especially fierce repression, conceal penises in their armpits. Desire and desiring women, terrifying to begin with, in the horror film are rendered monstrous.

However, I would argue that the film needs to be read at another level than that allowed by the logic of repression and displacement—or rather, read at a number of levels, with psychoanalysis allowing us access to only one. In a throwaway scene in *Rabid,* a female plastic surgery patient, returning to the clinic for yet another nose job that will better suit the specifications of her "daddy," waves a copy of Ernest Jones's *The Life and Work of Sigmund Freud.* "I'm terrified to find out what it really means," she says with an apologetic laugh, as the camera zooms in on the cover. Here the film seems to mock its psychoanalytically inclined interpreter, with a warning that the standard Freudian reading of oedipality, of the traumatic discovery of sexual difference and its repressive effects, will only take us so far.[12]

Freudian psychoanalysis allows for only an either/or option, and for a narrative that will flow from the realization of one of these options: it's either a penis or a vagina, or much the same thing, either a penis or an empty-site-of-lacking-the-penis.[13] Any overcharged image, in dream or text, must inevitably be identified as symbolically one or the other. When the symbol appears out of place—a penis in a woman's armpit, for example—the either/or logic is not yet disrupted, though one is alerted to a disturbance in the psychosexual field, signifying excessive dread or impossible desire (neurosis, perversion, fetishism) that can only reveal itself in distorted fashion. In *Rabid,* this would lead us to the reading mentioned above, where an insistence on the phallic-ness of Rose's armpit-penis commits us to a textual exegesis whose primary task is to account for the presence of this phallus-out-of-place, this female phallus signifying both plenitude (libidinal excess) and powerlessness (castration).

However, one is still puzzled by the seemingly gratuitous siting of the penis in the armpit. The first displacement (penis to woman) is one we're comfortably familiar with, and well-trained to read, but why the second displacement to this modestly erogenous, hugely undertheorized zone? Keeping in mind Cronenberg's warning about the inefficacy of Freudian interpretation, I would suggest that this second displacement constitutes something of an absurdist detail, denoting not an excess of dread (which would instantiate a second or chain of displacements), but the insufficiency of genital iconography as master-key to the film. In other words, *Rabid* is not "just another" story of a phallic woman, but a film that takes the horrific convention of the phallic woman as its starting point, in order to imagine a sexual economy—a posthuman sexual economy, if you will—founded on sexual indeterminacy rather than sexual difference.

Rabid, like *Alien,* works to multiply instances of sexual indeterminacy, so that sexual indeterminacy becomes the grounds for (rather than an effect of) the text. One can pursue this argument by examining the status of the victims in both films. At the most basic level, traditional gender roles are consistently inverted, in a sexual economy where males as well as females may be "raped," or even impregnated, as in the case of *Alien*'s Kane: the anatomical indeterminacy of the predator becomes the condition of the prey as well. But more significant than simple inversion is the sheer inconsequentiality of sexual difference within the "desiring" economy each film sets out. The predator need take no account of the usual orifices set out for sexual reception: the Alien plants its embryo through mouth and throat into the gastrointestinal system; Rose's mutant organ makes an opening on whatever part of her victim's body is proximate. Since the genital identity of the victim means nothing, any convenient body is penetrable and thus desirable; desire can run along any channel so long as there's a warm body at the end of it. In other words, desire and thus narrative in both films are governed by a logic of metonymy,[14] directed merely by contiguity and accident, and taking no account of the "proper" channels laid out by sexual difference. The proper object of desire is the next one who walks by.

Rabid establishes virtually no "characterization" of a pre-phallic Rose (who doesn't even speak until she wakes up from her postoperation coma, a month after the wreck) that could account for her as the catalyst for this troublesome and unmediated dynamic of wanting, immediately, what one sees. Here one must look again to the narrative logic of contingency. A series of accidents, rapidly unfolded within the first ten minutes of the film, leads up to Rose's transformation: the accident of the collision; the accident of having her upper body burnt when the motorcycle lands on it; the accident of crashing near the plastic surgery clinic where Dr. Keloid, restive under the practice of cosmetic medicine, is itching to try out his new skin grafting technique; the accident, perhaps, of cancer, which develops and spreads according to a logic that eludes us. (Keloid acknowledges that cancer might result from the internal grafts—a possible though insufficient explanation for Rose's mutation.) In this case, within a medical rather than sexual discursive framework, metonymic narrative enacts the vicissitudes and fragilities of the human body, prone to disfigurement, alteration, diseases for which there is no known cure.[15]

However, the human body in *Rabid* is not merely fragile. It is also unexpectedly, imaginatively generative. Keloid's experiment involves the removal of healthy thigh skin tissue, which before grafting is rendered "morphogenetically neutral" to "lose its specificity as both thigh tissue and skin tissue." His expectation is that the neutralized tissue will regenerate as, for example, intestinal tissue as well as abdominal skin, picking up cues from undamaged cells in the abdominal area: "neutral field tissue has the same ability to form any part of the human body that the tissue of a human embryo has." But embryonic cells, with this astounding metamorphic ability, sometimes pick up the wrong signals and generate monsters, and Rose's neutralized tissue—human and not-human, Rose and not-Rose—reinvents itself and Rose's body along with it. Her gastrointestinal system is damaged; her neutralized tissue reshapes her as one who does not need to eat (indeed, she cannot eat—food makes her vomit), but can absorb nutrients only from human blood. And her neutralized tissue generates an organ to obtain that blood. The human body, deprived of its usual mechanisms, draws up a new blueprint for survival and creates new ones.

A crucial point here is that sexuality is only one of several registers within which the concept of "the human" is problematized in *Rabid.* The film emphasizes the metamorphic, unstable quality of bodily identity, setting forth a spectacle of a body reconfiguring itself according to an unexpected internal logic. This logic is derived "accidentally" to be sure, but it's a rigorous and economical logic, and one invented by the body itself while Rose slumbers through her month-long coma.

Rose's "phallus," within this account, is much more than a phallus: taking it literally for the moment, it's rather an organ notable for its elegant efficiency in drawing the blood Rose needs to survive. Rose's "phallus" resembles a hypodermic needle as much as a penis, and her "pleasure" entails the introjection rather than the ejaculation of bodily fluids. Experimental medicine has transformed Rose into not only a sexual freak (a sort of armpit hermaphrodite), but also a cyborgic entity. The sign of the mechanical disrupts—or rather is simultaneous with—the sign of the biological/sexual, so that the image cannot be interpreted according to the "either/or" logic of anatomical difference. Instead, the image builds up a superfluity of (not necessarily consistent) meanings through the logic of "and . . . and . . . and." More simply, and to use the terminology of the Freud I have sometimes been maligning, the image is overdetermined (phallus, hypodermic needle, digestive apparatus), and the narrative that circulates around that image is overdetermined as well, both functioning simultaneously within several discursive fields.

The joke the film plays on the viewer is that, though it negotiates so smoothly within the discursive field of sexuality, it isn't much concerned with sex (at least, sex as we know it). Contrast Cronenberg's earlier *They Came from Within* (1975), which also uses the device of medical experimentation initiating a narrative of metonymic desire. In this case, a doctor has engineered a slug-like parasite that blocks the psychic mechanisms inhibiting human libido. The parasite's host leaps on the most proximate human body in a sexual frenzy and transmits both slug and libidinal ecstasy; the "disease" spreads through an island-bound apartment complex and eventually, as the credits roll, to Montreal. One of the film's victims is infected in the laundry room, where she's a spectre of grim, asexual efficiency. She later appears at her apartment door, her face smeared with makeup, to attack a waiter delivering a meal. "I'm hungry," she croons. "I'm hungry for lo-ove!" Food is rejected in favor of sex. In *Rabid,* by comparison, one of Rose's victims appears at a diner and snatches a barbecued chicken from a truck driver. "Gotta eat! gotta eat!" he cries, tearing at the chicken; when that doesn't satisfy his hunger he grabs the arm of the waitress and gnaws at that instead. Food is rejected here too, and the body is desired, ravenously so—but desired as food, not for sex.

Or perhaps it would be better to speak of economies of wanting, of appetite and drive, rather than economies of desire in *Rabid.* Desire suggests the transformation of drive through such mechanisms as repression and displacement, fantasy and cathexis; the idiosyncratic logics of the unconscious dictate object-choice, perverse or otherwise. In *Rabid* (as in *They Came from Within*), there is no need to reroute drive through the unconscious, no need for cathexis. Every object-choice (within the human field) is a good choice. Biting people, which should represent a perverse transfiguration of the sexual drive, is "appropriate" behavior within this economy of wanting.

Seemingly a film about perverse sexuality, *Rabid* allows room for neither perversion nor sexuality. Drives other than sexual ones dictate intimate relations between human beings. It's not that "traditional" desire is absent from the film: Rose's mere presence is able to elicit in men an immediate and wistful longing, manifesting itself in behaviors ranging from protectiveness to predatory lechery. Rose plays the game, presenting as a passive, available, nubile woman, allowing herself to be picked up in porno theaters, shopping malls, and lobbies. None of her would-be predators realize until too late that they're trapped in a new narrative. The spectre of the most traditional desire of all—heterosexuality, one anatomical configuration seeking its opposite term—haunts the film as a sort of outmoded relic in this new posthuman economy of wanting.

Alien Embodiments

> "Hey—sure wouldn't mind gettin' some more of that Arcturian poontang."
> "Yeah, but the one you had was a *male*!"
> "Doesn't matter when it's Arcturian."
> —Dinner conversation among male marines, *Aliens*

Alien's crew sets down on a bleak, inhospitable planetoid whose atmosphere, as the Science Officer Ash notes reverently, is "almost primordial." Lambert, Kane, and Dallas, fitted up in bulky and unwieldy suits, struggle across the surface of the planet through howling winds to track the source of the distress signal—a derelict space ship of nonhuman design. In the cockpit of the ship they discover the skeleton of its gigantic and vaguely elephantine alien pilot, whose outbent ribs indicate it has died by implosion, and, in the bowels of the ship, the egg chamber of the Alien species that killed the derelict's crew. Within this chamber, into which Kane is lowered with a rope, all is suffocatingly hot and damp—Kane reports that "It's like the goddam tropics in here." Innumerable eggs, the size of swollen melons and with an unpleasant gray-pink pulpy skin, rest on a muddy and glutinous surface, like the floor of a bog rather than a spaceship. When Kane's motion alerts the larval Alien within one egg to the presence of prey, the egg's distinctly vaginal opening,[16] coated with slime, unfolds to reveal a moist half-globe of throbbing tissue, faintly reminiscent of a cervix. Kane incautiously thrusts his head over the top of the egg to get a better look at the stirrings within, and the larval Alien leaps from its egg, breaks through Kane's helmet, and attaches itself to his face. The film cuts to show Lambert and Dallas dragging Kane's comatose body back to the tug.

Thus far the film may be said to offer a narrative of human beings, buttressed by an impressive (but ultimately, insufficient) technology, confronting primordial forces of nature they hope to control, but cannot even begin to comprehend. If one works within a (human) sexual economy of reading—where the visible signs of reproduction, its dampness and viscosity, serve as markers of "the feminine"—one might code "Nature" as feminine, and technology (by default or by tradition) masculine. The film, in fact, seems to demand such a reading. The derelict ship, viewed from the outside, appears as a rough compendium of recognizably human reproductive organs. At first viewing, the body of the ship is hidden behind a rock outcropping; all one sees are two gigantic, unmistakably phallic extensions of the ship thrusting out into space. At the center of the ship, between these outthrust phalluses, are three orifices looking like stylized vaginas; the three humans climbing through one of these outsized openings appear as insignificantly puny, and seem to be embarked on that all-too-familiar, always unpleasant, voyage into the body of the Monstrous Mother. And indeed, the inner walls of the derelict appear as the interior of a body, constructed of a material that could be metal, silicon, or bone, and covered with a repetitive pattern that could be drawn from an intestinal or skeletal system.

And yet, the interior of the derelict might be said to resemble the stuff lying under the hood of a car just as well as an organic body, human or otherwise. The film invites one into a sheerly psychoanalytic reading by strewing the field with the usual images, phalluses and vaginas turning up in every direction one turns, but then shifts into a field of discourse for which psychoanalytic interpretation is quite useless. Is the dead helmscreature masculine or feminine? How do these elephantine aliens reproduce? Do they experience sexual desire? Such questions are blindingly irrelevant—at least, the text does not allow space for their relevance. The spectacle presented by the dead elephantalien (to my mind the most eerie and beautiful moment in the film) opens up speculative itineraries that lead one away from considerations of sexual difference and sexual identity, into the imaginative realm of alternate embodiments.

The pilot lies recumbent on a control chair, its head poised just below a massive telescope—or rather, the pilot seems to have grown out the chair, or grown into it. H. R. Giger writes that the pilot was "conceived as one of my biomechanoids, attached to the seat so as to form a single unit" (34). The grooves and ridges of the chair blend imperceptibly into the pilot's desiccated body, so that the alien seems to be made of metal or silicon, and the chair of bone. This alien's scale dwarfs the human scale: while the alien is huge, so much so that the humans seem incongruously tiny beside it, the cockpit is nonetheless enormous on a scale disproportionate to the alien's size; the cavernous proportions of the cockpit are emphasized by the vast amounts of empty space between the control chair and the cockpit walls, bespeaking an aesthetic sensibility and economy of design that contrasts jarringly with the claustrophobically small and crowded control room of the *Nostromo.*

Briefly, the camera frames Dallas's face in juxtaposition with the helmscreature's as if to emphasize the contrast between the human, fully biological entity suited clumsily into a mechanism that enables it to breathe the alien air, and the biomechanoid, cybernetic entity whose strange and beautiful body impossibly confounds the (human) distinction between organism and machine. But this contrast is a misleading one. On the one hand, the film works to erode the boundary between human and machine, particularly via the liminal figure of Ash, the android successfully masquerading as a twitchy, anal-retentive human. On the other hand, the pathos of the two juxtaposed faces—Dallas's marked with nervousness and bewilderment, the helmscreature's frozen in a death mask of agony and astonishment—points toward their common status, despite their overwhelmingly different bodily forms, as mere prey for the superior Alien. "The human," within the terms of evolution and species identity, represents no apex of perfection. Both elephantalien and human, both complex sentient species, are evolutionarily "unfit" in contrast with the Alien—the probably but not certainly sentient organism that Ash will later admire for its "structural perfection."

This is one of several moves the film makes (despite the ninth-inning heroics of Ripley)[17] in disallowing "the human" as a recuperable category. The Alien is not interested in humans for all their special wonderfulness; rather, the human species is one of any number it can parasite in order to reproduce its own. And while the first *Alien* film is never as clear on this as the third, it allows for the possibility that Alien parasitism entails the absorption of selective (or random) characteristics of the host species, so that the creature birthed from Kane is an undifferentiated conglomerate of human, elephantalien, and countless other species forms. Within these terms, "the human" serves as just one bodily possibility within the infinite repertoire of the metamorphic Alien. The Alien lacks its own discrete and proper form. It represents an alternate species logic, a logic of adaptation through transformation across generations. Human bodily identity, by contrast, seems inefficient and monotonously repetitive.

And as a crucial part of its alternate logic of embodiment, the Alien offers us a spectacle of reproduction and birth outside of a human sexual economy. We are on one level invited to read the Alien symbolically, within the terms of human sexuality as determined by genital identity. The image is confused—the Alien is overdetermined within the sexual field, legible as both phallic and vaginal, male and female.[18] But too insistent an attempt to frame the Alien within this field obscures the fact that the Alien—taken literally for the moment—has no genitals at all. Here, as in *Rabid,* the excess and confusion of genital iconography point toward the insufficiency of human sexual difference as the key to the deciphering of horrific embodiment. The Alien does not reproduce by means of heterosexual coupling, and it bypasses the reproductive organs of its host, so that human males can be "mothers" just as easily as females. The Alien has no unconscious or super-ego, human psychic formations that allegedly result from the attempt to master the trauma of sexual difference: Ash describes it as lacking "conscience, remorse, or delusions of morality."

The important difference in the film is not "the oldest difference in the world," between men and women,[19] but the difference between Alien and not-Alien (with the human subsumed into this latter category).

In discussing the ramifications of this difference, I would like first to consider the sheer alien-ness of the Alien. Viewing the larval Alien from above, as it sits on Kane's face, one sees a bony ridge or spine that continues into a very long hard tail (wrapped around Kane's neck); on either side of the ridge is a pulpy gray-beige body covering Kane's cheeks and eyes; and at the top of the body, at each side, are four long, almost skeletal fingers, uncannily human in their tripartite division and perfect fingernails. The larval Alien resembles a spider, it resembles a crustacean, it resembles a human hand pushed over Kane's face in a suffocating embrace; in its confusion of tentatively recognizable forms it resembles nothing at all. The adult Alien is even more unplaceable with taxonomies of natural history: a living organism composed of materials that should be inimical to organic life (cells made of "polarized siliçon" and "molecular acid" for blood). Vaguely humanoid in form (at times, since the flexible body collapses and contorts in unexpected ways), the adult Alien also resembles a reptile, an insect, and a full endoskeleton stripped of flesh. Its shape calls to mind both the interior of the derelict and the biomechanoid form of the elephantalien: the Alien does not use machinery except for camouflage, blending perfectly with the black innards of the *Nostromo,* but it has become machine, or rather become cybernite, an entity simultaneously organic and inorganic.

Compare Stephen Prince's discussion of Carpenter's *The Thing:* "As a horribly anomalous animal, the thing represents a form of cosmic pollution, an entity existing outside the accepted categories that give shape to human life and knowledge" (26). Like the Thing, the Alien constitutes a collapsing of multiple and incompatible morphic possibilities into one amorphous embodiment—a logic of "identity" that serves as an alternative, or possibly a ontological challenge, to a human one predicated on a body that's a discrete, bounded, and stable unit. As indeterminate phenomenon, however, the Alien does not just serve as a contrast to the human. Alien amorphousness "infects" the human as, on the one hand, the integral human body is penetrated, invaded, rendered no-longer-human through Alien impregnation, and on the other, the Alien samples and absorbs human characteristics along with anything else it can find, so that the "human species" is no longer viable as an integral and distinct category.

In other words, while I am arguing for a reading that will attend to the alien-ness of the Alien, I am also arguing for a reading that collapses the distinction between Alien and human—but only on the terms of allowing for "the human" as an imploded category. The film accomplishes this implosion on many levels, not the least of which is defamiliarizing the human body by eviscerating it. In the famous dinner table scene where the Alien bursts from Kane's chest, the human subject is turned inside out and revealed as nothing but blood and guts, nothing but a body composed of matter as viscous and horrific as the slime that drips from the Alien's mouth. In the equally shocking scene where Ash is decapitated, the human subject is everted and revealed as nothing but circuitry and tubing. The scene's effectiveness depends partly on the spectator's surprise: though we've been given plenty of reasons to distrust Ash, until this moment we have had no cause to suppose him an android. The human body is exposed as merely a replicable construct, an entity that can be produced within a technical rather than sexual economy; and it is further exposed as a liminal form, a fully mechanical construct that so beautifully duplicates "human" appearance and behavior that one can't tell the difference.

But the scene is also fully effective as a moment of body horror. Ash's oddly fleshy viscera are every bit as disgusting as Kane's, coated in a thick, milky fluid that also spurts and bubbles from Ash's mouth. The scene renders visible the difference between android and human, but in emphasizing the gross corporeality of both android and human body, it also works to level that difference. More startlingly, the scene catches Ash in a fit of what looks like hysterics, matching Ripley's hysterics after she discovers the *Nostromo* has been set up by the Company. Ripley slaps Ash around, sobbing and shouting incoherently; when she gets her self-control back it's at the cost of an hysterical nose bleed. A shot of the single red drop trickling from her nose is countered with a shot of a single white drop trickling down Ash's forehead, and then Ash starts to giggle. In the ensuing violence where Ripley is nearly killed and Ash dismantled by Parker and Lambert, Ash moves first lumberingly and then spasmodically, hands and face shaking, arms flailing wildly as he emits high-pitched squealing noises. Presumably the android has malfunctioned, but the uncanniness of the scene can be traced to the near-perfect resemblance between mechanical malfunction and human hysterics, as if hysteria were not the result of the intricate interplay between human unconscious and human body, but merely a spectacle to be acted out across any body at all.

Alien, like body horror in general, works to disallow human specificity at every level, to evacuate the "human subject" in terms of bodily, species, sexual, and psychological identity. What body horror offers in place of this is a human/not-human subject, a posthuman subject, speculations on alternate logics of identity that rupture and exceed the ones we know. That such alternate identities are usually presented in the "negative" register of horror—that the ruination of the human subject is accomplished graphically and violently—need not, I think, argue for the sadism of the genre and its fans. It might argue instead for a genre without nostalgia, without investment in "the human" as a discrete and stable category, and for a viewer whose pleasure derives from spectacular enactments of the posthuman rather than through mechanisms of identification. As Jacques Derrida writes of deconstructions in general, even those who want to look often "turn their eyes away when faced by the as yet unnamable which is proclaiming itself and which can do so, as is necessary whenever a birth is in the offing, only under the species of the nonspecies, in the formless, mute, infant, and terrifying form of monstrosity" (293). If posthumanity is our postmodern condition, we will find this "as yet unnamable" state bodied forth most fully in the monster film.

Notes

1. Some critics posit that horror has been so transformed that new generic labelings are called for. Philip Brophy proposes "Horrality," a neologism that collapses "horror, textuality, morality, [and] hilarity" (3), and is designed to call attention to recent horror's playful, gleefully excessive manipulation of horrific conventions, its "violent awareness of itself as a saturated genre" (5). In a more disapproving vein, Harvey R. Greenberg proposes " 'cruel' or 'hardcore' horror, comprising pictures that skirt over the edge of the impermissible visually and psychologically, or plunge over the edge" (85).
2. See Robin Wood 17–22 and Greenberg 85–87.
3. Critics who have aided my attempts to imagine alternatives to psychoideological readings of horror film include Tania Modleski, who examines the postmodern narrative strategies of horror film; Brophy, particularly in his attention to special effects technologies; Robert Wood, who argues that genre hybridization accounts for the discursive overdetermination of a film like *Alien;* and Stephen Prince, who proposes a social anthropological theory of horror based on Douglas's and Leach's work on liminality within symbolic systems. Though I do not cite them elsewhere, my speculations on posthuman embodiment have been influenced by Donna Haraway and Judith Butler.

   I would also like to thank Judith Halberstam, Mary Klages, Ira Livingston, F. Tyler Stevens, and Eric White for their suggestions and conversation about body horror.
4. My discussion throughout is indebted to Julia Kristeva, who proposes abjection—a state of bodily and psychic liminality where the subject is not-subject—as a paradoxical status that simultaneously repels and fascinates, that elicits nausea but also inspires speculation and creativity.
5. I have devoted some time to unfolding Wood's psychoideological analysis of horror not because his essay constitutes the last word on the subject, but because he is so frequently cited by later critics. Perhaps more than anyone else,

Wood has set the stage for an expectation that the horror film advance a "progressive" political agenda, so that films that fail (are read as failing) to do so are dismissed as morbid, conscienceless, or downright pernicious.

6. See, for example, Barbara Creed 53, 65, and Peter Fitting 285–86, 291. Prince, though he breaks deliberately from the practice of the psychoanalysis of horror, also posits a "conserving" function for the genre: by violating the "distinctions that give form to the social environment" and to human identity, horror film ultimately reaffirms them (28).
7. Throughout this article, when referring to the film's creature I have capitalized "Alien," to differentiate from other uses of the word "alien."
8. Creed's is the most powerful and extended argument along these lines. Working from Kristeva, Creed argues in general that horror enables its spectators to affirm themselves as subjects within the symbolic order by confronting and then refusing an undifferentiated corporeal identity associated with the Mother.
9. See Fitting 287–89, Jeff Gould 282–83, Greenberg 97–103.
10. See Robin Wood 27–28, and James H. Kavanaugh and Judith Newton. My quick summary above does not do justice to the complex and rigorous argumentation of these latter two.
11. See especially Creed. As several critics have pointed out, the sequel *Aliens,* directed by James Cameron, insistently places Alien-ness within the register of femininity, with its climactic battle between the good mother Ripley and the Queen mother of all Aliens (Constance Penley 124–25, Robert Wood 8).
12. As Ira Livingston puts it, "through her [the Freud-reading patient] the film singles out for special punishment its would-be interpreters" (530). My reading throughout is indebted to Livingston's multi-discursive analysis of *Rabid.*
13. My argument here has been shaped by discussion with Mary Klages, as well as by Luce Irigaray's witty satire of Freud in her essay "This Sex Which Is Not One."
14. Modleski, working from the assumption that metonymy is generally "the principle by which narrative is constructed," argues that metonymy in *Rabid* "becomes the means by which narrative is *disordered,* revealing a view of the world in which the center no longer holds" (161). Here I would prefer to distinguish between a narrative ordered by other tropic devices as well as metonymy, and a narrative that proceeds purely on the basis of contingency; and to characterize as differently ordered rather than disordered this narrative that "spreads," like infection, from bodily unit to bodily unit.
15. See Boss for a discussion of horror film in the context of disease and modern medicine.
16. H. R. Giger, the Swiss artist who designed most of the alien creatures and effects for the film, describes his first model of the egg thusly: "When I take off the plastic cloths in which my work is draped, there is a howl of laughter from the whole group. I had lovingly endowed this egg with an inner and

outer vulva. To make it all look more organic, I filled some more preservatives with clay and arranged these semi-transparent little sausages on the pink aperture." The egg orifice was later doubled like a cross, to reduce the absolute resemblance to the female genitals and achieve the effect of a "flower opening," lest a too anatomically correct design (albeit adorned with sausages) invite "trouble, especially in Catholic countries" (46).

17. While many critics see *Alien*'s conclusion as recuperative of a humanist ideology—the lone heroic individual defeats the Alien and saves herself and the cat—I am not inclined to give the ending such weight. At best, I would argue, Ripley's heroics (or confused and desperate attempts to survive) are available, perhaps nostalgically, as one strategy in the repertoire of posthuman identities and behaviors posited by the film.

18. The field is further confused by more than the sheer excess of sexual signs. In some scenes, the Alien's lips pull back to reveal nothing but the two sets of teeth—it's all *dentata* and no *vagina.* The inner "phallic" mouth, an attenuated oblong box whose metallic white latticework is clearly visible under the slime, appears simultaneously organic and inorganic, like the hypodermic penis in *Rabid.*

19. Several critics have noted the fact that there is no sexual chemistry between male and female crew members on the *Nostromo;* some find it refreshing (Newton 84), some disappointing (Robin Wood 27). Penley writes that "Dan O'Bannon's treatment for the first film was unique in writing each role to be played by either a man or a woman" (124–25). In other words, the script in general works to minimize the "oldest difference in the world."

Works Cited

Butler, Judith. *Gender Trouble: Feminism and the Subversion of Identity.* New York: Routledge, 1990.

Creed, Barbara. "Horror and the Monstrous-Feminine: An Imaginary Abjection." *Screen* 27 (1986): 44–70.

Derrida, Jacques. "Structure, Sign, and Play in the Discourse of the Human Sciences." *Writing and Difference.* New York: Routledge, 1978. 278–93.

Elkins, Charles, ed. "Symposium on *Alien.*" *Science-Fiction Studies* 7.3 (1980): 278–304.

Fitting, Peter. "The Second Alien." Elkins 285–93.

Giger, H. R. *Giger's Alien.* Beverly Hills: Morpheus International, 1979.

Gould, Jeff. "The Destruction of the Social by the Organic in *Alien.*" Elkins 282–85.

Greenberg, Harvey R., M. D. "Reimagining the Gargoyle: Psychoanalytic Notes on *Alien.*" *Close Encounters: Film, Feminism, and Science Fiction.* Ed. Constance Penley et al. Minneapolis: University of Minnesota Press, 1991. 83–103.

Haraway, Donna J. "A Cyborg Manifesto: Science, Technology, and Socialist-Feminism in the Late Twentieth Century." *Simians, Cyborgs, and Women: The Reinvention of Nature.* New York: Routledge, 1991. 149–81.

Irigaray, Luce. "This Sex Which Is Not One." *This Sex Which Is Not One.* Ithaca: Cornell University Press, 1985. 23–33.

Kavanaugh, James H. "Feminism, Humanism and Science in *Alien.*" Kuhn 73–81.

Kristeva, Julia. *Powers of Horror: An Essay on Abjection.* New York: Columbia University Press, 1982.

Kuhn, Annette, ed. *Alien Zone: Cultural Theory and Contemporary Science Fiction Cinema.* New York: Verso, 1990.

Livingston, Ira. "The Traffic in Leeches: David Cronenberg's *Rabid* and the Semiotics of Parasitism." *American Imago* 50.4 (1993): 515–33.

Modleski, Tania. "The Terror of Pleasure: The Contemporary Horror Film and Postmodern Theory." *Studies in Entertainment: Critical Approaches to Mass Culture.* Ed. Tania Modleski. Bloomington: Indiana University Press, 1986. 155–66.

Newton, Judith. "Feminism and Anxiety in *Alien.*" Kuhn 82–87.

Penley, Constance. "Time Travel, Primal Scene and the Critical Dystopia." Kuhn 116–27.

Prince, Stephen. "Dread, Taboo and *The Thing:* Toward a Social Theory of the Horror Film." *Wide Angle* 10 (1988): 19–29.

Wood, Robert E. "Cross Talk: The Implications of Generic Hybridization in the *Alien* Films." *Studies in the Humanities* 15 (1988): 1–12.

Wood, Robin. "Introduction." *American Nightmare: Essays on the Horror Film.* Ed. Andrew Britton et al. Toronto: Festival of Festivals, 1979. 7–28.

ELEVEN

Terminating Bodies: Toward a Cyborg History of Abortion

Carol Mason

> Andrew Ross [posing a question to Donna Haraway]: It seems clear that there are good cyborgs and there are bad cyborgs, and that the cyborg itself is a contested location. . . . How do you prevent, or how do you think about ways of preventing, cyborgism from being a myth that can swing both ways, especially when the picture of cyborg social relations that you present is so fractured and volatile and bereft of guarantees? (*Technoculture* 7)

Andrew Ross pinpoints a key problem when he recognizes "cyborgism" as a "myth that can swing both ways." Referring to Haraway's essay, "A Manifesto for Cyborgs: Science, Technology, and Socialist-Feminism in the 1980s,"[1] Ross tries to draw out the possible dangers involved in promoting an ironic myth that can, precisely because of the nature of irony, serve the rhetorical needs of two opposing political stances. The cyborg, a creature of interdependent cybernetic and organic elements, was introduced by Haraway to the academic community in 1985 as a playfully self-conscious attempt to argue "for *pleasure* in the confusion of boundaries [separating human from machine, human from animal, bodies from technology, female from male] and *responsibility* in their construction" (150). But Ross's question reminds those of us enamored of Haraway's cyborg that pleasure's twin is danger.

What are the dangers involved in embracing the cyborg as a myth that aims, in Haraway's words, to "transform the despised metaphors of both organic and technological vision [and] to foreground specific positioning, multiple mediation, partial perspective, and therefore [to serve as] as possible allegory for antiracist feminist scientific and political knowledge?"[2] One of the dangers is getting caught up in judging the "good cyborgs" and the "bad cyborgs" and forgetting that, as Ross says, "the cyborg itself is a contested location." I see this happening, for example, in feminist science fiction novels like Marge Piercy's *He, She, and It* (1991), in which even the good cyborg intensifies and re-solidifies identities and boundaries marked by sexual difference.[3] The practice of identifying good and bad cyborgs often reifies political identities and social relations as individual bodies. James Cameron's 1991 film *Terminator 2* (*T2*) also encourages viewers to engage in this practice, to choose which cyborgs are good and bad, and which social relations are to be saved or not. Subtitled "Judgment Day," the film clearly invites us to play spectator-god, to distinguish the good from the bad, to salvage relations like the nuclear family, and to ignore Ross's point that "the cyborg itself is a contested location."

I resist reading *T2* according to good/bad cyborgs and look instead at the "cyborg social relations"—or "cyborgism"—embodied by characters. Although Ross treats both approaches as one, I think tracing "cyborgism" is a less dangerous reading practice than locating good and bad cyborgs. I prefer tracing the cyborgism of a situation over locating good and bad cyborgs because it focuses on the historical and discursive interplay among bodies rather than on the bodies themselves. Furthermore, at least in the case of *T2*, locating good and bad cyborgs is counterproductive to reading the cyborg social relations at work in the film. The project of "cyborgism," my analysis will show, tends to result in the disembodiment of individual subjects and the disaggregation of historical social relations that comprise a subject. In other words, cyborgism as a reading practice reveals how subjectivities are made and remade—how they are reproduced.

Haraway promotes this practice more than the project of judging good and bad cyborgs. She asks, "why should our bodies end at the skin?" (178) and suggests that we can analyze political situations according to the poststructuralist revelations that cyborgs afford us. Two of those revelations are that organic unity (or bodily integrity) is itself a myth, and that technophobia leads people to surrender some very important tools to an elite group (corporate/military scientists) known for their eugenic ways. Haraway knows that relinquishing the ideal of organic unity or a suspicion of technology is dangerous. She admits this danger more clearly in the interview with Ross than she does in the manifesto, where it may have seemed antithetical. Yet Haraway's devotion to her ironic myth prevails in her response to Ross's question about cyborgism's dangerous capacity to "swing both ways":

> Donna Haraway: Well, I guess I just think it [cyborgism] *is* bereft of secure guarantees. And to some degree, it's a refusal to give away the game, even though we're not entering it on unequal terms. It is entirely possible, even likely, that people who want to make cyborg social realities and images to be more contested places—where people have different kinds of say about the shape of their lives—will lose, and are losing all over the world. One would be a fool, I think, to ignore that. However, that doesn't mean we have to give up the game, cash in our chips, and go home. I think that those are the places where we need to keep contesting. It's like refusing to give away the notion of democracy to the right wing in the United States. (*Technoculture* 7–8)

My response to this call "to keep contesting" begins with the distinction between identifying good cyborgs from bad cyborgs and recognizing the "cyborgism" of a situation. My reading of *T2* illustrates this distinction and explores how bodies do and don't "end at the skin." On one hand, my reading emphasizes race, which appears to be located "at the skin." The "fact of blackness"[4] is the most immutable and taken-for-granted signifier of individual embodiment in *T2*, a movie that otherwise flaunts stunning changes in personal appearance and revels in bodily plasticity. On the other hand, my reading will illustrate how even the fact of blackness in *T2* is historically and discursively produced by and producing a fact of whiteness, if you will. In this way, *T2* urges us to think in terms other than individual bodies, or "singularities," as our editors suggest.

Gilles Deleuze and Felix Guattari also explore ways in which bodies can be seen as non-unitary, heterogenous machines that may emphasize rather than obscure the contingent and ongoing process of reproducing subjectivities. They best describe bodies as machines in discussing how a bumblebee is part of the reproductive system of a red clover, or how the male wasp is part of the reproductive system of the orchid which "attracts and intercepts [it] by carrying on its flower the image and the odor of the female wasp" (*Anti-Oedipus* 285). According to Deleuze and Guattari, the bee and clover or the wasp and orchid are, together, a complicated reproductive machine. Moreover, they tell us, "we are misled by considering any complicated machine as a single thing." The same is true, I will argue, in *Terminator 2:* the truly complicated machinery in Cameron's film is not a terminator; it is the confrontation between Miles Dyson, the father of Skynet (an automated defense program that leads to cyborg revolution) and Sarah Connor, the mother of John, a boy who is destined to lead humans in future battles against the cyborgs. Dyson, who is black, and Connor, who is white, are like the bee and clover; they serve as parts of each other's "reproductive" systems. I will argue that this Connor-Dyson opposition (re)produces culturally recognizable subjectivities that need to be read not merely as stereotypes or "monolithic ideological effects" (Goldberg 181). Connor and Dyson aren't merely representations or one-on-one denotations of white and black, female and male, worker and professional, or "mannish lesbian" and "noble negro"; they work together as a reproductive machine lubricated by these historical residues. Only with a sense of history (and specifically a history of eugenics, lynching, and population and reproductive control) can we detect and understand *how* this cultural "reproduction" occurs.

Following my analysis of *T2*, then, will be a brief defense of history as a safeguard against a cyborgism that swings both ways, and an application of this safeguarding, which will contextualize cyborgism in the arena of abortion politics. This will allow us to bring into question Haraway's cyborg as a myth with "no origin story" (150), and to disturb the dichotomies upon which the abortion debate is founded.

T2 is most obviously about a good cyborg (the Terminator played by Arnold Schwarzenegger) and a bad one (the T-1000 played by Robert Patrick). These cyborgs arrive from the postapocalyptic future in which John Connor leads a human rebellion against the cyborg anarchy, which began when an automated military defense system, Skynet, came to consciousness and attempted nuclear genocide of human beings. This confusing time loop allows the cyborgs to travel into the past to prevent (the Terminator's mission) or to ensure (T-1000's mission) Skynet's cyborg reign. For the T-1000, this means killing John Connor while he's still a child, before he grows up to be a military leader against cyborgs. The Terminator, in direct conflict, must protect John. Thus we have our good/bad cyborg theme.

At the opening of the film, however, viewers aren't sure who's good and who's bad. Since this is a sequel to a film in which Schwarzenegger played the evil terminator, viewers expect him to be bad again. But unlike his shapeshifting nemesis (T-1000), whose bodily shape and appearance can change, the Terminator's character can change and, in fact, is constantly changing, not only from bad to good or from monstrous to humane (if not human). The Terminator also traverses a continuum of masculinities.

According to Elizabeth Traube, mass culture's attempt to remasculinize America after the Vietnam and Watergate era was not a re-establishment of any one type of masculinity. Rather, she says, different masculinities range from "working-class resentments of managerial control" (20), as portrayed by *Rambo,* to "good, nurturing fathers [that serve as] substitutes for bad, overambitious mothers" (25) of the professional-managerial class (as we see in *Kramer vs. Kramer* and *Mr. Mom*). The Terminator traverses that continuum of masculinities, moving from the mass-murdering type to the sensitive man who learns the value of crying. It is tempting to argue that Schwarzenegger in *T2* in fact contains all these masculinities as he neutralizes the "schizophrenic" conflict between horror movie dads (Jack Nicholson in *The Shining*) and family melodrama dads (Dustin Hoffman in *Kramer vs. Kramer*).[5]

But I agree with other scholars whose work indicates that the Terminator's masculinity is constantly being defined, re-produced, or (Traube's term) re-invented. Jonathan Goldberg, for example, convinces me that the Terminator's masculinity is produced and reproduced through the tensions and interplay between "leathersexual" images reminiscent of gay male porn and the taken-for-granted "(hetero-)sexual allure" of Schwarzenegger's physique. To this insistence that the Terminator has both a "straight masculinity" and a queer one, Traube might add that the Terminator has both a working-class masculinity (as is evident in his Ramboesque irreverence for the law) and a professional-managerial masculinity (he would, Sarah tells us, never hit John or get drunk but would be the "best father" like corporate advertising man Ted Kramer).

Of course, the Terminator isn't the only one whose sexuality is reinvented in *T2*. After scratching "no fate" in a table with a buckknife, Sarah Connor has a horrible dream in which she sees herself (as she appeared in the 1984 *Terminator*) and her son in a playground that is engulfed and disintegrated by a nuclear inferno. The dream sequence juxtaposes her butchy, militant and muscular, streamlined look with her big-haired, softer and rounder, femmey image from the first terminator film. While this is a reinscription of the fame and bisexual appeal of Schwarzenegger as a professional bodybuilder,[6] it is also an image of maternal love hardened by the spectre of nuclear war. Sarah Connor points a high-tech rifle at the origin of nuclear apocalypse, the intellectual father of the computer-automated military defense system called Skynet. She can focus in on the supposed cause: a black man named Dyson. Sarah Connor, once an unassuming working-class waitress, now has alliances in Mexico and Nicaragua, arms "ripped" with muscles, sophisticated guerrilla gear, and an anti-nuclear attitude.

As she approaches Miles Dyson, we see that he, in contrast, has achieved the American Dream of prosperity, complete with loving family. He has a swimming pool, a state-of-the-art home computer, and a boy who is playfully stalking him with a remote control truck, a toy version of what

John Connor and the Terminator use to race to the scene. It is the plastic battery-run truck—not that hulking old metal Ford—that saves Dyson by attacking his ankle, where he stoops and consequently misses the silent shot fired by Sarah. Sarah shifts to machine-gun mode and starts blasting, intending to kill Dyson regardless of the pleading child or his screaming wife. In this way, Sarah's attitude is "anti" nuclear family as well as against nuclear war. But she can't go through with the murder and is joined by the Terminator and her son John, whom she tearfully tells that she has always loved.

She verbalizes her reasons for attacking Dyson in terms that Alice Echols would attribute to a particular school of feminist thinking that polarizes and essentializes men's culture and women's culture. Connor says that nuclear holocaust is due to "fucking men like" Dyson who don't know what it means to "create a life," to feel something "living inside you," and continues to posit the production of the H-bomb as men's jealous response to the reproductive power that women embody. Thus she sets up a cultural opposition between men and women based on biological functions or the lack of them. These generalizations are cut short by son John who dismisses them as ironically not "constructive" just as mom Connor is accusing men of being innately "destructive."

This belittling of cultural feminism seems especially interesting given Connor's confrontation with Dyson, the head of an affluent black family, a black male intellectual, a scientific professional in the tradition of Dr. Cliff Huxtable of "The Cosby Show." When confronted with the fact that he's responsible for future nuclear apocalypse, Dyson agrees (at the prompting of his wife) to change things, destroy his research, and eventually sacrifices himself in a manner that might be described as noble. A confrontation which pits the noble black man against the protective white mother has historical roots that cross and interconnect. I want to examine the discursive web at work between Miles Dyson and Sarah Connor, and suggest that these historical interfacings are more convoluted and perhaps more lethal than the shapeshifting liquid metal terminator, the T-1000, or his nemesis, the more mechanical cyborg played by Schwarzenegger.

Sarah Connor's attempt to murder Miles Dyson puts in opposition a white woman and a black man, whom she calls "motherfucker." This opposition structurally resembles a historical scenario dating from the nineteenth century and used as a rationale for lynching. According to many African American women intellectuals,[7] and to quote Jane Gaines, there is an ostensible opposition between black men and white women that is "a sexual scenario to rival the Oedipal myth" (24).

In the nineteenth century white men accused black men of raping white women as a reason for lynching and/or castrating former slaves. The displacement of the reality of white men (slaveholders) victimizing black women onto a fiction of black men victimizing white women is still at work in the late twentieth century. I see that displacing logic in action when fears about *genocide* get presented as something that black men do to white women instead of what white men do to black women. That displacing logic of lynching culture is replayed in *T2*, but not in terms of rape. As Connor screams at Dyson while condemning the designers of thermonuclear warheads, the movie presents a counter-historical reversal in that a white woman is agonized by genocidal technologies produced by a black man. Let me first discuss how Dyson is seen as the threat of genocide. I will then discuss how, historically, it's white men who threaten women of color with genocide, not vice versa as *T2* suggests.

Instead of being lynched or castrated for raping, Dyson is castrated and murdered for supposedly masterminding genocide, which would amount to annihilating all future generations. Most straightforwardly, Connor sets out to kill Dyson because he is the one responsible for producing the technology that leads to Skynet, the defense system that comes to consciousness and overrides human authority to create the nuclear-apocalyptic "Judgment Day." But Sarah's quest to stop Skynet's creator, Dyson, is a personal necessity, not a political opposition; it is an act of transcendent maternal love. Sarah Connor illustrates Jonathan Schell's understanding of "disarmament as an act of parental love" (as quoted in Sofia 58). Moreover, Schell redefines parenting in an all-inclusive way: "Nuclear peril makes all of us, whether we happen to have children of our own or not, the parents of all future generations" (Sofia 58). Given this notion of genocide as the extermination of "all future generations," the creator of Skynet is a genocidal threat in a very specific way. The creator of Skynet is a sort of cosmic abortionist, which the movie's script insinuates when things go awry in the Cyberdyne lab and Miles Dyson says, "we have to abort."

This link of abortion to nuclear genocide is made in the first terminator movie, where Reese, a soldier from the future, is sent back in time to protect Sarah, the future mother of John Connor, from the Terminator. Reese explains to a psychiatrist how the Terminator has gone back in time to thwart the resistance movement led by adult John Connor in the twenty-first century (and thereby ensure the success of Skynet's genocidal apocalypse) by killing John Connor's mother before he is conceived. The patronizing psychiatrist sums up this transtemporal effort as an attempt at "a sort of a retroactive abortion." In addition to possibly invoking "primal scene" fantasies,[8] the time-travel that Reese speaks of in this scene is the convention of a "time loop": a trope that structures not only classic science fiction but the rhetoric surrounding abortion as well.

The time-loop trope that makes sense out of a "retroactive abortion" (an attempt to kill an adult man before he is conceived) is the same logic used to argue against terminating a pregnancy. In anti-abortion rhetoric, the fetus represents both a future and former child. It simultaneously denotes the embryo "you once were" and the child that embryo may become. This sense of a "collapsed future"[9] is exemplified in fetal rights legislation, too. For instance, a 1978 Missouri law, *Berstrasser v. Mitchell,* ruled that

> a child born with brain damage because of injury inflicted to his mother's uterus during a caesarian section for prior delivery had an independent cause of action against the physicians who performed the surgery before the child's conception. (Poovey 248)

The same back-to-the-future logic that grants this child the right to sue doctors, to punish them for the future brain damage that they supposedly caused before that brain-damaged child is conceived (and has a brain to be damaged) is the same science fiction logic that compels the Terminator of the first film to go back in time to "terminate" John Connor by killing his mom before she conceives him. This time-loop logic is fully established and taken for granted from the beginning of *Terminator 2,* when both the Terminator and the T-1000 come back to the past to ensure or prevent a future that is both a foregone conclusion and a fundamental premise.

Therefore, when Dyson is set up as the man who both has already destroyed and has yet to destroy humanity, it makes sense in a screwy but culturally valid way. When Dyson proposes that they "have to abort," we know that he's talking about aborting humanity as well as aborting the plan to stop Skynet; that the nuclear apocalypse that has already happened in the future is imminent; and that time loops allow all these "retroactive abortion[s]" to occur.

While these links may not be explicitly played out in the terminator movies, the merging of these discourses on/of science fiction, thermonuclear militarism, and abortion politics succeeds in ideologically saturating a confrontation between a black man and a white woman. Also in this toxic discursive confluence is, I contend, the displacing logic of lynching culture which swaps a reality of white men victimizing black women for a fiction of black men victimizing white women. Historically, it has been women of color who have suffered the genocidal technologies designed and legislated by white men, not white women at the mercy of black men, as *Terminator 2* suggests in opposing Connor and Dyson. Based on research presented by Angela Davis in "Racism, Birth Control and Reproductive Rights," the following is a chronological smattering of the genocidal technologies at work in twentieth-century USA alone. I offer this to shed historical light on the Connor-Dyson opposition in general and Connor's cultural-feminist comment in particular.

In the first two decades of the century, according to Davis, fears of "race suicide"—of the white race killing itself off—compelled leaders like Teddy Roosevelt, the Carnegies, and the Kelloggs to support eugenics programs. By 1930, 26 states passed compulsory sterilization laws and thousands of "unfit" people of color were surgically prevented from reproducing. In 1939, a national birth control organization developed a "Negro Project" in which black ministers were insidiously recruited to lead local birth control committees. According to Davis, "this episode in the birth control movement confirmed the ideological victory of the racism associated with eugenic ideas" (215).

In the '40s and the '50s, "birth control" became "population control." Franklin Roosevelt considered "overpopulation" the cause of Puerto Rico's economic difficulty and launched an experimental campaign there, which resulted in a 20 percent decline in population growth by the mid-1960s. By the 1970s, more than 35 percent of all Puerto Rican women of child-bearing age were surgically sterilized. In 1972, 16,000 women and 8,000 men were sterilized in the U.S. according to the Department of Health, Education, and Welfare, which later estimated that between 100,000 and 200,000 sterilizations had actually been funded that year by the federal government. These figures, disproportionately, represented black and Chicana women. The same government system that offered free sterilization either (1) wouldn't pay for abortion services, or (2) provided services on the condition that the patient agreed to sterilization. For example, Davis reports, an Indian Health Services Hospital in Oklahoma was reported to sterilize 1 out of every 4 Native American women giving birth in that federal facility.

It would be inaccurate to attribute these fearful attempts to prevent white "race suicide" to white men alone. Not only were there individual women in the eugenics movements—birth control leader and feminist Margaret Sanger was an administrator of the aforementioned "Negro Project"—but collectively, white feminist interests opposed attempts to stop compulsory sterilization. According to Thomas Shapiro, in 1972 Planned Parenthood, NOW, and the National Association to Repeal Abortion Laws (NARAL) protested a sterilization guideline proposed by the Committee to End Sterilization Abuse (CESA, the predecessor of CARASA, Committee for Abortion Rights and Against Sterilization Abuse). The guideline called for a 30-day waiting period, which addressed the needs of the undereducated or non-English speaking women who are most susceptible to sterilization abuse (Shapiro 140). This liberal feminist protest stemmed either from a class/race bias that prevented white feminists from seeing the advantage of the guideline, or from a fear that the concomitant abortion cause would be hurt if they aligned themselves with this other, seemingly unrelated issue.

There is a common denominator between this abridged history of feminist involvement in "genocidal technologies" and Sarah Connor's cultural feminist outburst that ignores racial inequities while bemoaning gender differences. In both cases, the issue of racism is ignored and even advanced by a "woman's issue." I think the same historical dynamic of producing racism through women's rights is happening when white mom Sarah Connor, who exalts her ability to "create a living thing" while condemning men's potential for destruction, blames a black man for the death of humanity, the abortion of all future generations.[10]

It is, then, a historically convoluted and ideologically saturated scene when Connor and all she represents—white womanhood, cultural feminism, the nuclear mother for whom "disarmament is an act of parental love"—confronts Dyson as the mythic black menace opposed to white womanhood, the apocalyptic "motherfucker" that Cameron's narrative via Sarah Connor contends he is, and as the cosmic abortionist of all future generations. At work here are many potent technologies that are far more elusive and contortionistic than the T-1000, and far more intricate and insidious than technology as represented by the computer chip and the cyborgian arm that Connor and company must retrieve from Cyberdyne labs and destroy.

Dyson takes them to Cyberdyne when it is made clear to him who he's dealing with. After being hit by a bullet meant to kill him, Dyson asks, "Who are you people?" In response, John, a little white boy, hands a knife over to the big white man and says severely, "Show them." The Terminator proceeds to terrorize Dyson and his wife by cutting off the cyber-skin to reveal a mechanical arm and hand. Narratively, the implicit answer to "who are you people" is: we're the people who have arms just like the one Dyson has in a glass dome at Cyberdyne labs. Dyson recognizes the Terminator's arm because, as we learn when Dyson takes them into the labs, he has been studying a similar appendage, a left-over arm from the previous Terminator, whose story comprised the first *Terminator* film (1984).

Historically, however, the implicit answer to Dyson's question is: we're the people who cut flesh without flinching. Seeing a white man with a knife evokes not a Freudian castration anxiety but rather a historical castration anxiety that I mentioned earlier with regard to Gaines's "scenario." Like Gaines, bell hooks contends that "practically all the visual images that remain of lynchings of black males by white mobs show blacks to be sexually mutilated, usually castrated" (81). So when John triumphantly grabs that chip and arm at Cyberdyne, it is no unloaded exclamation he makes: "We've got Skynet by the balls now!" The castration of Skynet—taking Skynet "by the balls"—is a substitution for the lynching/castration/murder of Dyson, who is the symbolic black "motherfucker" who serves as the scapegoat for the actual men who conceive thermonuclear war[11] and so many other technologies of genocide.

And what, according to *Terminator 2*, sets off this network of genocidal technologies? Again, Miles Dyson sets it off. I've already argued that he does so "ideologically," as the mythic black menace opposed to white womanhood. But also, visually and literally, Dyson is the detonator. His body, immobilized and holding the ironic "remote control" device aimed at the explosives he's sitting beside, mechanically gives out. The remote control technology represented by his son's little truck that saved him at home can't save him now. As his last, sharp intakes of breath gradually slow like a clock or a toy that needs rewinding, his arm falls and triggers the explosives. It's tempting here to argue that Miles Dyson is the man-machine combination—the "real" cyborg as opposed to either terminator in this movie.

My point about Dyson, however, is that (t)his body *as opposed by* Connor's is historically and narratively part of a tremendous network of technologies that include government policies on eugenics; feminist stances on abortion; intersecting fictions of race, masculinity, and femininity; the science of nuclear war; and the institutions of rape and lynching. Dyson "embodies" these technologies only in relation to and in as much as Connor. While we make meaning of all that Dyson embodies, we are also making meaning of what Connor embodies. The black, upwardly mobile intellectual father is produced with and through the white militant feminist mother, and vice versa. Yet my project is not to flesh out explicitly this particular historical and discursive production of white American femininity and black American masculinity.

Instead, I'm suggesting that with a sense of history as something as contingent and constructed as a cyborg, we can situate political problems in relations between or among bodies instead of positing politics in the body "itself." These relations can't be contained in individual or singular bodies—regardless of their modern, medico-psychological depth or their postmodern, spectacular, schizophrenic depthlessness—because the discourses that (re)produce those bodies are historical. To say, then, as does Haraway (181), that we "know of no other time in history when there was greater need for political unity to confront effectively the dominations of 'race,' 'gender,' 'sexuality,' and 'class' " is to view history myopically. To situate a call for a posthuman "political unity" (or collectivity based on "affinity") primarily with regard to the "high-tech" late twentieth century is to reduce "posthumanity" or anti-humanism to an experience.[12] Such a reduction misses the opportunity to claim "posthumanity" as an essentially political condition or (even!) platform which, despite postmodern time-looped distortions of temporality, can also be historicized.

One way of creating that history is to remember how "the announcement that 'the body' has a *history*" is coterminous with the announcement that the modern self or the Enlightenment subject is also a historical convention. There is much slippage between these ideas of "the body" and "the self" as historical, hence denaturalized and political "constructs." For example, arguments surrounding abortion in the 1960s, '70s and '80s reflect how radical feminists' interrogation of the political *subject* gave way to liberal feminists' protection of the political *body.* This slippage from considering political subjects to considering political bodies is not exclusive to feminists; often the intellectual move from political subjects to political bodies conflates the two, so that when we talk about "our bodies" we talk about "our selves." Not enough attention has been paid to this easy movement between bodies and selves; and only by historicizing the emergence of these ideas can we see what opportunities and dangers are at stake in this slippage.

Therefore, history may be "inefficient as a method of processing information," as our editors insist (see introduction), but it is not inadequate or obsolete. Halberstam and Livingston are more explicit than Haraway in describing why they privilege this era over "any other time in history"; they focus on the present as an unprecedented time or condition when "history, social history, chronological history, are dying with the white male of western metaphysics and consequently it is no longer enough to say where we have been." Although I share with the editors the poststructuralist aim to detect and eradicate metaphysics,[13] I fear that "we" will stop examining where we've been or that "we" replace rather than enhance that examination with a new notion or representation of our bodies, our selves. I fear that without saying "where we've been" and without defining specific political goals, even the best intentions to theorize embodiment as something other than "fixed location in a reified body" (Haraway 195) can obscure the historical and discursive production of subjectivities, and consequently hide some of the political opportunities *and* pitfalls available in understanding such productions. Defining or owning up to what those political opportunities and pitfalls are is what can prevent the posthuman project from becoming too much like *Terminator 2,* where the razzle-dazzle of single, plastic (not to mention white and male) bodies gets more airtime than the implied or direct, historical and discursive confrontations that supply the plasticity out of which culturally valid bodies and subjectivities arise.

Thus far, I have explored the cyborg as a representative theory of political embodiment that often obscures the historical and discursive (re)-production of subjects. I have argued that by jettisoning our reliance on what the editors call "singularities" we can understand not only that the Terminator embodies many different masculinities, but that those multiple masculinities become culturally coherent and humanified by way of conflicts between Dyson and Connor. Together and in conjunction with the Terminator, Dyson and Connor re-produce queerness, race, gender, and class not according to separate or singular bodies but according to historical discourses. I hope this reading of *Terminator 2* has provided an example of what thinking in terms other than "singularities" can yield, and has demonstrated that it's really not cyborgs, or any individual bodies, that we need to examine. It's the examination of contingent and perpetual process of historical and discursive re-production that can allow us to better locate, articulate, and specify the aims of this "political unity" or "posthuman" "we."

To close, I want to show how a focus on cyborgism rather than good or bad cyborgs can enable the specifying of political aims. This entails concretizing a theoretical situation in the specific political case of abortion. Mary Poovey has written a proposal against regarding the abortion controversy as a fight over individual rights and bodily integrity. In "The Abortion Question and the Death of Man," Poovey offers what she hopes "will be an extremely controversial argument" that begins with an examination of the "metaphysics of substance that is implicit in the discourse of rights [and] historically related to the basic tenets of individualism" (243). Poovey anticipates controversy because she proposes not only an alternative line of argument but also an attack on these "cherished ideals." Basically, she suggests that we conceptualize the individual in terms of its "heterogeneity or nonunitary nature" and situate it in a network of social relations. To do this, she argues, we must dispense with the body as a requisite for legal personhood.

It would appear as if Poovey is applying a new cyborgistic reading to the old case of abortion. Like Haraway, she is calling for a sense of responsibility in the construction of social boundaries. Like my *T2* argument, her analysis ends up disembodying the forces involved to reveal the historical and discursive web that ensnares women who wish to terminate a pregnancy. Thus she can speculate that this "nonindividualistic politics" could put abortion "alongside other services that recognize social needs—

> such services as prenatal care, child day-care for working parents, and medical care for those unable to care for themselves. Far from making the abortion issue more arcane or difficult to identify with,

> this repositioning of abortion within the landscape of contemporary issues might well increase the number of people willing to support abortion on demand, for it would align advocates of safe and legal abortions with the millions of women and men who support safe and effective birth control and with the growing numbers of people who endorse plans for day care and parental leave, as well as some system of subsidized health care capable of guaranteeing affordable medical service to everyone. (252–53)

In effect, she's looking to conduct a cyborgistic legislation for people affected by the abortion "debate," and to strategize according to the disaggregation of social relations that such a reading reveals.

It interests me greatly, then, to discover that her interpretive and political strategy does not stem from any leftist manifesto or anything desiring to be associated with terms like feminism or socialism. Poovey notifies her readers up front:

> Ironically, the terms of the alternative politics I will outline here have already been introduced into the abortion debate by its most conservative participants—those people who endorse the idea of fetal personhood. The fetal personhood argument, after all, makes it clear that the embodied individual is only one of many possible interpretations of what counts as a legal person possessed of rights. This position therefore introduced the possibility that legal personhood might be assigned to some unit that is lesser or greater than the embodied individual. (251)

The irony Poovey reveals here brings us back to the interview question I opened with. Ross's observation that the cyborg can "swing both ways" clashes with Haraway's claim that cyborgs "have no truck with bisexuality" (150). What cyborgs lack in bi*sex*uality, Ross might argue, they make up for with bi*text*uality. Poovey's analysis indicates that cyborgism has indeed swung both ways textually, or at least contextually, since the same reading practice that creates a disaggregation of the discourses of the body is compelling in both the socialist feminist context and the pro-life legal context. Moreover, this irony Poovey discovers suggests that Haraway's model of "social relations" (i.e., the cyborg) is not what this strategy has swung from but swung to. In other words, the conservative right's conception of fetal rights has, in its disembodiment of legal personhood, predated if not anticipated the political strategy that is Haraway's cyborgism. Here we have an opportunity to question whether cyborgs really are, as Haraway claims, "exceedingly *unfaithful* to their origins" (151, my emphasis). To push the point, I might go so far as to suggest that the fetus is the "original" cyborg, in that the sociolegal myth of prenatal life is a cyborg construction exceedingly *faithful* to its origins.

In any case, Poovey's work helps to reveal the magnitude of "the main trouble with cyborgs" which, according to Haraway, "is that they are the illegitimate offspring of militarism and patriarchal capitalism," to which they should be "unfaithful." In other words, cyborgs are made of the stuff they are meant to be disloyal to, to disrespect, and to dismantle.[14] It is important to recognize the extent to which this "trouble" can be exploited. Putting the cyborg's bitextuality (its ability to swing both ways) in the context of abortion shows how disturbing this "trouble" can be. Specifically, it can disturb the dichotomies on which the abortion debate rests.

Without succumbing to the pleasure/danger of squaring off "good cyborgs" from bad—which occurs at the cost of engaging cyborgism as an analytic of discursive histories—we can take this opportunity to acknowledge that, at least with regard to abortion, the relationship between bodily integrity and rights is produced and sustained equally by two dichotomized camps, namely pro-choice and pro-life. Poovey reveals how pro-choice philosophy is as much a producer as an opponent of fetal rights thinking when she discusses *Roe v. Wade*:

> the basis for shifting the discussion about abortion to the issue of fetal rights was laid not by *Webster* but by *Roe v. Wade*. For, in locating the "point in time" at which an individual acquires rights at a moment *before* birth through the concepts of viability and "potential" life, *Roe* implicitly called attention to the arbitrariness of relying on the biological state of embodiment for a definition of the social concept of "meaningful life" and, by extension, the sociolegal concept of legal personhood. (248)

Thus Poovey begins to demonstrate how the pro-choice movement, whose paragon of rights is *Roe*, and the pro-life movement, which owes the legal articulation of "life" to *Roe*, are discursively intertwined and mutually producing the terms of the abortion debate. Like Dyson and Connor, pro-life and pro-choice work together as complicated discursive machinery which repeatedly churns out subjectivities embodied as pro-choicers and pro-lifers—subjects as real, for example, as terrorist Paul Hill, who assassinated abortionist John Britton in Pensacola in July 1994.

While there is no doubt that we can recognize Paul Hill as an intolerably "bad cyborg," we must also recognize the historical residues that lubricate his delusional gears. "Fight this [abortion] as you would fight slavery!" Hill yelled at the camera, implying that fetus is to woman as slave is to slave master, which, in the American context, racializes both the fetus as black and the woman as white. Hill thus reproduces abortion as an implicit contest between races.[15]

We know better than to simply substitute the pro-choice/pro-life debate for another two-party opposition like black against white or Dyson against Connor. Too often such oppositions, represented as single debates or embodied by singular individuals, swing both ways. "Cyborgism" as a political reading practice—if not reduced to demonizing bad "cyborgs" and exalting good ones—can collapse oppositions. Tracing the cyborgism of abortion politics in the USA exposes and explodes the false dichotomies on which the abortion debate is hinged. This leads us, as Poovey notes, with a different set of political alliances and aims, while enabling, as Haraway says, a refusal to "give up the game," a refusal to accept or reproduce (in terms of "choice") the right's articulation of the problem of abortion.

Notes

I want to thank those who helped with this essay which, although not collaboratively written, is a product of many discussions with many folks, including: Steph Athey, Maureen Konkle, Betty Joseph, John Mowitt, Thea Petchler, Paula Rabinowitz, Marty Roth, and the audience and members of the Science Fiction panel at the 1992 Midwest Modern Language Association meeting, where a version of this paper was presented. For good criticism, I thank the editors of and readers for this volume.

1. The essay first appeared in *Socialist Review* 80 (1985): 65–108. All quotations and pagination I use, unless noted, refer to the reprinted version, published in Haraway's *Simians, Cyborgs, and Women: The Reinvention of Nature* (1991).
2. This quotation appears in "The Actors Are Cyborg, Nature Is Coyote, and the Geography Is Elsewhere: Postscript to 'Cyborgs at Large' " in *Technoculture,* 21.
3. Claudia Springer discusses the widespread tendency of cyborgs in popular culture to intensify rather than eliminate gender differences. See her "The Pleasure of the Interface," *Screen* 32:3 (Autumn 1991): 303–23. Marge Piercy's novel *He, She, and It* neutralizes, contains, and embodies any shifting definitions of gender. Mother and father figures serve to oedipalize Piercy's cyborg, which results in naturalizing and heterosexualizing Yod. Yod has all the traditional attributes of a man because a human male, Avram, designs him for defense, equipping him with cybernetic muscles and penis, and with software that gives him pleasure in violence. As high-tech midwife to Avram's creation, Malkah programs emotions in Yod, designs some sort of surge protector that gives him an easy coming to consciousness, and serves as his sexual teacher; she is the source for "feminine" traits. Piercy's cultural feminism

is evident here; in positing a figure who clearly is intended to subvert "natural" sexual difference, she re-establishes that difference as the fundamental problem. Yod is, dialectically enough, a cyborg who synthesizes the best of both sexed worlds and as such never interrogates the categories of masculinity and femininity but contains and confirms them.

4. This phrase and concept is Frantz Fanon's. See chapter 5 of *Black Skin, White Masks* (New York: Grove Press, 1967).
5. Vivian Sobchack discusses this "schizophrenic" relationship (11) between movie genres in the '80s in "Child/Alien/Father: Patriarchal Crisis and Generic Exchange" in *Close Encounters: Film, Feminism, and Science Fiction.* Constance Penley, Elisabeth Lyon, Lynn Spigel, Janet Bergstrom, eds. (Minneapolis: University of Minnesota Press, 1991).
6. In addition to Goldberg's argument, novelist Harry Crews also maintains that Arnold Schwarzenegger has a bisexual appeal: "All [Schwarzenegger] had to do was take off his shirt to be worth ten million dollars because every man's dick in the audience got hard and every woman got wet to the knees." *Body* (New York: Poseidon Press, 1990) 74–75.
7. The works of nineteenth-century writers Pauline Hopkins, Ida B. Wells, and Anna Julia Cooper are the basis for contemporary understandings of lynching culture. See Angela Davis, "Racism and the Myth of the Black Rapist," *Woman, Race and Class* (New York: Random House, 1983); Jane Gaines, "White Privilege and Looking Relations," *Screen* 29 (Autumn 1988): 12–27; Hazel Carby, " 'On the Threshold of Women's Era': Lynching, Empire, and Sexuality in Black Feminist Theory," *"Race," Writing, and Difference,* ed. Henry Louis Gates, Jr. (Chicago: University of Chicago Press), 1986; bell hooks, *Black Looks* (Boston: South End Press), 1992. See also Stephanie Athey, *Contested Bodies: The Writing of Whiteness and Gender in American Literature* (DAI, 1993).
8. Constance Penley argues this in "Time Travel, Primal Scene, and the Critical Dystopia," *Camera Obscura* 15 (1986): 67–84.
9. Like the Star Child of Kubrick's 1967 *2001: A Space Odyssey,* the terminator and Reese embody what Zoe Sofia calls a "collapsed future." Like the terminator who enters the past naked and in "the fetal position," the aged astronaut of *2001* becomes a cosmic fetus that is both futuristic and embryonic (Sofia 57). As Sofia explains, anti-abortion ideology thrives on this concept of a collapsed future by insisting that not only did you "come from" but that you "once were" an embryo, and by promoting the fetus as "an astronaut in an interuterine space ship" (56–57). This second tactic is especially rich given the decision of the New Right to launch in the early 1980s an anti-abortion campaign as "part of a general conservative strategy to reprivatize health and welfare services while freeing up more resources for arms build-up," including Reagan's—but not George Lucas's—Star Wars (Sofia 54).
10. The historical dynamic of racism and women's rights that I nod to here is the subject of an emerging approach to feminist studies and the history of

women's literary production. For an in-depth discussion of this dynamic as it resonates with my argument, see *Contested Bodies: The Writing of Whiteness and Gender in American Literature,* a recent dissertation by Stephanie Athey.

11. This is not to say that there are no blacks who support SDI, that blacks are collectively opposed to genocidal technologies like thermonuclear warheads (in fact one African American nuclear engineer was awarded a special citation from the Secretary of Defense for his work on the Manhattan Project). Rather, my point lies in the displacing logic of lynching culture, the logic that produces "the myth of the black rapist" (as Davis calls it) which by no means maintains that there are no black men who rape.
12. Thanks to Thea Petchler for this insight.
13. This definition of poststructuralism's "official mission" is Fredric Jameson's (see "Actually Existing Marxism," *Polygraph* 6/7 1993, 170).
14. The root of this "trouble" and the source of the cyborg's irony which allows it to swing both ways may have already been located by Joan Scott. She suggests that Haraway's manifesto tautologically launches a critique of socialist feminism from within the philosophical confines of socialist feminism. See "Commentary: Cyborgian Socialists?" in *Coming to Terms,* Elizabeth Weed, ed. (New York: Routledge, 1989).
15. Hill's comparison of abortion to slavery is in the same rhetorical tradition that presents *Roe v. Wade* (1973) as homologous to *Dred Scott v. Sandford* (1857), the legislation that supposedly denied citizenship to African Americans on the basis of their being an "inferior class of people." I explore this homology and the issue of abortion as an implicit contest between races more fully in another essay, "Managing King and Administrating Life: Overturning Sovereign Power as a Recurrent Symbolic Narrative in American Politics" (presented at the national meeting of the American Political Science Association, September 1994).

Works Cited

Cameron, James, dir. *The Terminator,* 1984; *Aliens,* 1986; *Terminator 2,* 1991.

Davis, Angela Y. *Women, Race, and Class.* New York: Vintage, 1981.

Deleuze, Gilles, and Felix Guattari. *Anti-Oedipus: Capitalism and Schizophrenia.* Minneapolis: University of Minnesota Press, 1983.

Echols, Alice. "The Taming of the Id: Feminist Sexual Politics 1968–83." *Pleasure and Danger: Exploring Female Sexuality.* Carole Vance, ed. New York: Routledge, 1984.

Gaines, Jane. "White Privilege and Looking Relations: Race and Gender in Feminist Film Theory." *Screen* 29 (Autumn 1988): 12–27.

Goldberg, Jonathan. "Recalling Totalities: The Mirrored Stages of Arnold Schwarzenegger." *Differences* 4.1 (1992): 172–204.

Haraway, Donna. *Simians, Cyborgs, and Women: The Reinvention of Nature.* New York: Routledge, 1991.
hooks, bell. *Black Looks: Race and Representation.* Boston: South End Press, 1992.
Poovey, Mary. "The Abortion Question and the Death of Man." *Feminists Theorize the Political.* Joan W. Scott and Judith Butler, eds. New York: Routledge, 1992.
Ross, Andrew, and Constance Penley. *Technoculture.* Minneapolis: University of Minnesota Press, 1991.
Shapiro, Thomas. *Population Control Politics: Women, Sterilization, and Reproductive Choice.* Philadelphia: Temple University Press, 1985.
Sofia, Zoe. "Exterminating Fetuses: Abortion, Disarmament, and the Sexo-semiotics of Extraterrestrialism." *Diacritics* 14.2 (1984): 47–59.
Traube, Elizabeth. *Dreaming Identities: Class, Gender, and Generation in 1980s Hollywood Movies.* Boulder, San Francisco: Westview, 1992.

TWELVE

"Once They Were Men, Now They're Land Crabs": Monstrous Becomings in Evolutionist Cinema

Eric White

A defining characteristic of many of the monsters of Classical legend is their composite nature. Satyrs, gryphons, hydras, the sphinx, are traditionally classified as "monstrous" because their bodies mock the notion of organic unity in their arbitrary juxtaposition of anatomical features drawn from heterogeneous life-forms. Monsters are liminal entities that transgress the ontological categories and distinctions upon which the construction of a comprehensive and lasting representation of reality depends. Not surprisingly, encounters with such creatures typically evoke feelings of fear and loathing. But monsters may also produce astonishment and wonder in their beholder. The Ancients in fact often interpreted the bizarre singularity of monstrosity as a portent from beyond. Emissaries from worlds other than this one, from futures in which the seeming inevitability of the present shape of things has been rescinded in favor of an indeterminately immense range of fantastic possibilities, monsters can equally well inspire that vertiginous sensation of elation the Surrealists termed "the Marvellous."

In this chapter, I'm going to examine four exemplary film narratives in which the human body loses its specificity as "human" and becomes itself monstrously other. With reference to the evolutionary theory that in large measure inspired all four films, I intend to argue that such monstrous becomings can be understood to figure an evolutionist perspective on the human body as an assemblage of non-human parts. That is, evolutionist cinema renders the body monstrous by, so to speak, re-animating hitherto latent aspects of human nature, the genealogically prior forms of non-human life that constitute what I'd like to call the *menagerie within*. But the monstrous becomings of evolutionist cinema thematize not only this important if at times disturbing truth—namely, that "human nature" is *not* except as a monstrous amalgam of the non-human. In these films, the prospect of becoming even more monstrous than the present human form is ultimately affirmed, either openly or implicitly, as a destiny devoutly to be wished. To put it more generally, if these films could ever be said to credit the notion of a "providential design" inscribed in the natural world, that design would reside precisely in nature's capacity to produce *monstrosity.*

Becoming-Crustacean: *Attack of the Crab Monsters.* Directed by Roger Corman. Screenplay by Charles B. Griffith. USA: Los Altos Productions/Allied Artists, 1957.

—Martha Hunter, marine biologist: "Well, doctor?"
—Dr. Karl Weigand, nuclear physicist: "This is ridiculous. The molecular structure of the crab is entirely disrupted. There is no cohesion among the atoms."
—Hank Chapman, technician: "I don't understand."
—Weigand: "Neither do I. Apparently we have one of those biological freaks resulting from an overdose of radiation poisoning. The way to explain it is—look—electricity. The free electron in the copper atom breaks off to circle the next atom, taking the charge along the wire. Do you follow me, Hank?"
—Chapman: "I think so. The free electrons jump from atom to atom along the copper at the speed of light. I remember that from high school."
—Weigand: "Yes. Atom to atom. Well, something like that has happened to our crab. But instead of free electrons the crab has free atoms all disconnected. It's like a mass of liquid with a permanent shape. Any matter therefore that the crab eats will be assimilated in its body of solid energy, becoming part of the crab.
—Hunter: "Like the bodies of the dead men?"
—Weigand: "Yes. And their brain tissue, which after all is nothing more than a storage house for electrical impulses."
—Dale Drewer, terrestrial biologist: "That means that the crab can eat his victim's brain absorbing his mind intact and working."
—Weigand: "It's as good a theory as any other to explain what's happened."
—Hunter: "But Doctor, that theory doesn't explain why Jules' and Carson's minds have turned against us."
—Drewer: "Preservation of the species. Once they were men, now they're land crabs."

Perhaps the most memorable of the many "giant bug" films of the 1950s, Roger Corman's *Attack of the Crab Monsters* is distinctive not only for Corman's trademark tackiness and Absurdist panache but for the remarkably knowing manner in which the film plays evolutionism off against traditional myth. At the beginning of the film, a message scrolls across the screen advising the viewer that "you are about to land in a lonely zone of terror" wherein abound "frightening rumors about happenings way out beyond the laws of nature. . . . " These "happenings," unprecedented and unforeseen events that cannot be accounted for according to the known order of the natural world, are initially presented in the film as a sort of divine punishment. The film opens with footage of numerous hydrogen bomb explosions in the Pacific followed by scenes of immense tidal waves inundating seaside communities. An unidentified voice is then heard that seems to emanate from the clouds as it recites the passage in Genesis announcing the Flood: "And the LORD said, I will destroy man whom I have created from the face of the earth; both man, and beast, and the creeping thing, and the fowls of the air; for it repenteth me that I have made them" (Genesis 6:7). The wickedness of Man, as instanced specifically in the fantasies of apocalyptic omnipotence that accompanied the invention of nuclear weapons, is thus punished by a God who returns the world to a condition of primordial chaos, the traditional mythic emblem for which is the turbulent sea.

This recurrence of the flux of inchoate forms that obtained at the beginning of time entails as well an allusion to the Biblical story recounting the expulsion from the Garden of Eden. The Fall of Adam and Eve is updated in this case, however, as a Fall not merely into a material world of hardship, suffering, and mortality, but into the biological chaos of an evolutionary Nature. The turbulent fluidity of the primordial abyss finds its earthly avatar, in other words, in the phenomenon of biological evolution. Humanity is now compelled to inhabit a world in which the continuing emergence of unforeseen, hitherto unimaginable, and therefore monstrous life-forms disrupts every attempt to impose a satisfyingly human direction on the course of events. Henceforth, reality will transpire as a succession of accidents and unpredictable "happenings," the aleatory becoming of a relentlessly shapeshifting biosphere.

The chief representatives of evolutionary chaos in *Attack of the Crab Monsters* are common land crabs that have grown to house-size giants as a result of their exposure to intensely radioactive nuclear fall-out. The story unfolds as a life-and-death struggle between these evolutionary monstrosities and the team of research scientists sent to investigate the effects of radiation on the South Pacific island that is the crab's native habitat. As the team leader Dr. Weigand redundantly puts it, the crabs are "biological freaks resulting from an overdose of radiation poisoning." But their monstrosity resides not merely in their great size. When Weigand dissects the severed claw of one of the crab monsters he discovers that the crab's internal constitution is radically mutant: "The molecular structure of the crab is entirely disrupted. There is no cohesion among the atoms. . . . The crab has free atoms all disconnected. It's like a mass of liquid with a permanent shape." While the crab's outer form may enable the scientists to name the crab by consulting a taxonomic table of animal species, internally it has no fixed nature. The crab monster is, rather, a living incarnation of chaos itself.

Crabs are appropriately cast as embodiments of chaos. In her well-known discussion of "The Abominations of Leviticus," Mary Douglas observes that the dietary rules set forth in Leviticus generally rely on the following criterion to determine whether a given animal can be judged "clean" and thus fit to eat: the animal must come "equipped for the right kind of locomotion in its element." Crabs are consequently forbidden as unclean. Though the element proper to crustaceans is the sea, crabs possess legs that enable them to walk on the land. Inhabiting the margin between land and sea, with characteristics of both terrestrial and marine animals, crabs are in fact truly abominable creatures, "unholy," "polluted," and "monstrous." In their undecidability, they "confound the general scheme of the world" (Douglas, 1969: 55). Crabs serve notice, in other words, that the human species inhabits not a rational cosmos whose structure has been fixed for all time but an evolving multiplicity that will always become other than the present classificatory grid.

The giant land crabs of *Attack of the Crab Monsters* have as their express mission the destruction of the South Pacific island that is the setting for the story. They intend, that is, to make "make war on the world of men" by drowning the humanly intelligible universe in the turbulent flux of a chaotic pluriverse. *Attack of the Crab Monsters* thus stages a mythic combat between order and chaos that in fact transpires on a number of levels. First of all, the self-identical rational ego of Enlightenment thought finds itself threatened by a voraciously appetitive id. According to this view, the crabs are precisely id monsters whose hard chitinous exterior just barely manages to contain a surging internal flood of anarchic drives and instincts. As the crabs devour the various members of the research team one after another, the survivors are horrified to discover that the personalities of their former companions have not been extinguished but mysteriously live on inside the crabs. Before he is himself welcomed by his now crustacean colleagues into the "great common stomach" of one of the crabs, Weigand makes the point that because the crab he is examining "has free atoms all disconnected. . . . Any matter . . . that the crab eats will be assimilated in its body of solid energy," including human "brain tissue, which after all is nothing more than a storage house for electrical impulses." This means, Drewer concludes, that a crab can eat its "victim's brain absorbing his mind intact and working." The crabs then deploy these formerly autonomous and free-willing selves as lures with which to obtain further prey. When one of the crabs' victims summons a sleeping Martha Hunter—"Martha, Martha Hunter, awake Martha, awake, it is McLaine. Help me"—its use of the third person pronoun is telling: Hunter hears no authentic human voice but the call of a self that is merely an instrument or ruse by means of which a monstrous id pursues its own agenda. The surviving members of the research team are thus forced to concede their evolutionary kinship with other forms of life. "Man" loses his ontological privilege when it turns out that the putatively rational human mind operates as much in the service of unconscious biological instincts (like "preservation of the species") as any other creature.

In the course of making "war on the world of men," the crabs not only dissolve the putatively ontological distinctions between mind and body, human and animal. As might be expected (given the time-honored association of the "feminine" with chaos, materiality, the unconscious, and the body), the crabs also threaten the intelligibility of gender. It turns out that becoming-crustacean, at least for most of the crabs' victims in the film, entails as well becoming-female when they are captured and eaten by a female crab who adds their formerly "autonomous," "rational," "male" minds to her ever-expanding repertoire of selves. The crabs are as indifferent to the gender of their victims as they are unperturbed to find, after having feasted on numerous human bodies, that they will henceforth be creatures endowed with multiple personalities instead of a unitary self.

But giant female id monsters who deploy male egos as stratagems with which to secure further victims is not the only form female trouble takes in the film. The one female member of the research team is Martha Hunter. Significantly, she is a *marine* biologist and thus doubly associated already, by gender and professional specialization, with the forces of chaos. As the story unfolds, her commitment to her fiancé Dale Drewer begins to waver. It seems that Drewer has sublimated so much of his desiring energy into his doctoral research that he is now bereft of directly sexual desire. Thus frustrated, Hunter finds herself drawn to Hank Chapman, a technician with only a high school education who consequently lives less in his mind and more in his body. As she and Chapman prepare a trap for one of the crabs in a cave, she hints at her interest in him with this artful opener: "Lonesome in here. . . . You know I bet you could even be lonesome in a crowd. Unless, of course, you've found that special someone."

But before Hank has a chance to respond in kind, they are interrupted by the loud snoring of the sleeping crab, which shortly thereafter awakens and prevents the courtship from proceeding further.

Hunter and Chapman never do succeed in consummating their attraction for one another. At the end of the film, with the island now reduced by the crabs to a small rock outcropping upon which only a radio tower is left standing, the three surviving members of the research team—Martha, Dale, and Hank—have a climactic confrontation with what they hope is the lone surviving crab. When the crab attacks, Hank climbs the radio tower and brings it crashing down on top of the crab, electrocuting both the crab and himself in the process. Dale and Martha then embrace as Dale says "he gave his life" and Martha breathlessly responds, "I know." Martha's vagrant female desire is thus redeemed by Hank's heroic self-sacrifice. Henceforth, her desire will assume a properly spiritual character by finding its object in Dale Drewer, the highly sublimated man of science.

Of course, *Attack of the Crab Monsters* in no way asks its audience to take this concluding scene seriously. The ending of the film comes across as an extravagantly parodic replay of traditional myths of redemption from sinfulness. The Absurdist smirk the film wears throughout can in fact be said to affirm a precisely contrary point of view. That is to say, *Attack of the Crab Monsters* endorses not too covertly the prospect of becoming-crustacean. Prior to their being devoured, both Devereaux, the team botanist, and Carson, the geologist, have suffered serious accidents. Devereaux's hand has been amputated by a falling rock and Carson has broken a leg. It's not at all far fetched, I think, to interpret these injuries as symbolic "castrations," castrations suggestive of the repression of desire that is the prerequisite for successful sublimation according to the classic Freudian formulation. This symbolic castration no longer obtains once Devereaux has become-crustacean. Speaking to the other members of the research expedition, Jules the crab positively exults in his new mode of existence: "Something remarkable has happened to me. I want all of you to come and see for yourselves. . . . " When Drewer responds with a query about Carson, Carson's persona, which is now domiciled in the same crab, answers: "I'm here too. My leg no longer troubles me. It's almost exhilarating." Almost exhilarating indeed: becoming-crustacean lifts the censorship of the body habitually imposed by the repressive superego of the civilized subject. Though it is possible to amputate a crab claw, the loss is never permanent, as one crab smugly asserts: "So you have wounded me, and I must grow a new claw, well and good, for I can do it in a day." Phoenix-like, the body's desiring energy continually renews itself: crab monsters really do have more fun.

Becoming-Insect: *Five Million Years to Earth.* Directed by Roy Ward Baker. Screenplay by Nigel Kneale. UK: Hammer/Warner Brothers, 1967.

> You realize what you're implying: that we owe our human condition here to the intervention of insects? . . . So that's your great theory!

The recognition of humanity's evolutionary kinship with other life forms receives a far different treatment in *Five Million Years to Earth,* undoubtedly the finest of four films that British screenwriter Nigel Kneale wrote featuring his scientist-hero Bernard Quatermass. Instead of suggesting that becoming-animal amounts to an emancipation from the repressive constraints of civilization, in this film evolutionary kinship with "lower" forms of life is the occasion for Gothic horror. The discovery in the course of the film of human relatedness to other organisms entails a profoundly disquieting consequence. High intelligence, a trait often regarded as proof of human exceptionality, is shown to be intertwined with and frequently in the service of a quite "primitive" propensity for aggression, hierarchy, and xenophobia.

Five Million Years to Earth is in fact preoccupied not only with the evolutionary history of "human nature" but seeks as well to ascertain the origin of evil. This concern is signaled early on in the scene in which the audience is first introduced to Quatermass, who heads up Britain's civilian space agency. Quatermass is now informed by a government official that the civilian space program is about to be taken over by the military. As Quatermass listens in disgust, the military officer who will soon be his superior gives him a lecture on geopolitics: "the present world situation makes [it] quite clear" that "whoever plants [bases on the moon] first will be able to police the earth with ballistic missiles." From here, the film sets itself the task of tracing the genealogy of this militarist impulse.

The story told in *Five Million Years on Earth* goes something like this: the film opens at a London Underground station where workmen are tunneling an extension of an existing line. Here they discover the remains of some early hominids, "apemen," a momentous find that soon draws the attention of a paleontologist named Ronay whose team begins to excavate the site. The fossils show, Ronay explains to the press, that "creatures essentially resembling mankind walked this earth as long ago as five million years." The fossils are typical of hominids of the period except that their braincases are unusually large for such an early date. As excavation continues, the team discovers not only more fossils of early hominids but what at first appears to be a modern technological artifact, perhaps some sort of unexploded bomb or V-weapon from the war. An army demolitions team brought in to defuse the device is unable to identify it, but as they remove the clay in which it lies embedded they come upon more hominid fossils, including one that is *inside* the object. Ronay is struck by the inevitable implication: "Good heavens, that's no bomb! Whatever is it?" This is, of course, a rhetorical question, as both he and the viewer realize that the object—which has an odd insect-like appearance, like the carapace of a beetle—must in fact be a spacecraft, a spacecraft in which the apemen had apparently been passengers when it crashed five million years ago at a time when the Thames valley was still a primordial swamp.

Further investigation of the spaceship reveals a sealed-off compartment that mysteriously opens by itself to reveal the bodies of the ship's crew. Ronay, Quatermass and everyone else on hand are frightened and nauseated as they are confronted with the now rapidly decomposing bodies of what appear to be giant, three to four foot high locusts or grasshoppers. As Quatermass reads the paleontological team's test results—"weight and structure point to low gravity environment. A thin atmosphere"—Ronay speculates on the giant grasshoppers' likely place of origin: "Perhaps a world that's dead now, but a few million years ago could have been teeming with life." Quatermass catches the hint: "I wonder, a world that's been nearly worn out before anything turned up to claim it. Was this really a Martian?"

Quatermass then reports to the Minister of Defense that "these arthropods are not of this earth" and speculates that the Martians, realizing both that their own planet was dying and that they could not themselves adapt to the denser atmosphere and greater gravity of the Earth attempted a sort of proxy colonization by experimenting on the early hominids. The apemen were "altered by selective breeding, atomic surgery, methods we cannot guess. And returned with new faculties instilled in them, high intelligence, . . . perhaps something else." Ronay's assistant Barbara Judd sums up the matter succinctly: "as far as anybody is, we're the Martians now." The Minister of Defense, like most of the high government officials and military figures in the film, is not amused and contemptuously dismisses Quatermass: "You realize what you're implying, that we owe our human condition here to the intervention of insects! . . . So that's your great theory!"

But Quatermass's hypothesis regarding the monstrously composite character of the human form—part-simian, part-insect—is hardly controversial from the standpoint of contemporary evolutionary theory. Evolutionists repeatedly insist that evolution is a "conservationist." Lynn

Margulis and Dorion Sagan make just this point: "It is virtually an evolutionary principle that no body part or chemistry once crucial to an ancestor is ever lost without a trace, although such features can be radically altered. Life is extraordinarily conservative; organisms embody their own histories." Evolutionary biology thus describes the human body as emerging from a succession of instances of biological improvisation or bricolage. The body is not a perfectly resolved unity endowed with a unique and everlasting essence but an evolutionary makeshift, a historically contingent contrivance whose genealogical affiliation with every other kind of organism is manifest throughout. The human body, Margulis and Sagan say, is a "fleshy pastiche . . . of thousands of ancestral lives" (1991: 134, 14). The heterogeneous forms of life that have evolved over the past roughly 3.5 billion years have arisen less often from unprecedented innovation than from the rearrangement and amplification of a common store of possibilities. In that sense, the present human form is not an entity set apart from and against other forms of life but a provisional configuration of elements that appear ubiquitously in different combinations and with different emphases in the other species that populate the planet.

Lewis Thomas argues along similar lines: "a good case can be made for our nonexistence as entities. We are not made up, as we had always supposed, of successively enriched packets of our own parts." Instead, the body is better understood as a colony of symbiotically cohabiting elements beginning at the level of the cell itself. The eukaryotic (or nucleated) cell—the body's, as it were, elementary building block—is itself not even a unity. Its mitochondria "turn out to be little separate creatures, the colonial posterity of migrant prokaryocytes, probably primitive bacteria that swam into ancestral precursors of our eukaryotic cells and stayed there. Ever since, they have maintained themselves and their ways . . . with their own DNA and RNA quite different from ours." This composite character of the eukaryotic cell is replicated, moreover, at every higher level of organization: "It has been proposed that symbiotic linkages between prokaryotic cells were the origin of eukaryotes, and that fusion between different sorts of eukaryotes (e.g., motile, ciliated cells joined to phagocytic ones) led to the construction of the communities that eventually turned out to be metazoan creatures." The human body, Thomas goes on to say, is quite literally, then, an evolutionary *assemblage* (1974: 2, 8, 145).

The depiction in *Five Million Years to Earth* of the composite character of the body as an assemblage or pastiche of prior life forms is inflected according to the conventions of Gothic fiction. Humanity's arthropod heritage is not merely horrifying; it is, in fact, precisely satanic. Near the beginning of the film, Ronay's assistant Barbara Judd points out to Quatermass that the name of the street on which the tube station is located—Hobb's Lane—was formerly "Hob's" Lane; "Hob," she says, is an ancient name for the devil. Judd researches the history of the site and discovers that as recently as the 1920s and as far back as Roman times, Hobb's Lane has been associated with strange and infernal happenings, including the apparition of "hideous" humanoid figures resembling the five-million-year-old hominids whose fossils litter the site. Though Quatermass is inclined to doubt the supernatural—"No, my dear, we're both scientists. We simply can't pay regard to stuff like this"—he is eventually compelled to admit that the many stories of supernatural phenomena associated with Hobb's Lane must have a basis in fact: "I suppose it's possible for ghosts, let's use the word, to be phenomena that were badly observed and wrongly explained." When Quatermass enters the now fully excavated craft, he finds inscribed on the compartment housing the dead Martians, a pentacle, "one of the Cabalistic signs used in ancient magic." Moreover, Martian-like forms recur in human culture from Paleolithic cave art of shamans wearing antlered headdresses to medieval gargoyles and the horned god of satanic ritual, all of which, Quatermass supposes, are evidence of some unconscious "race memory."

But the truly satanic nature of the prehistoric Martian grasshoppers is not revealed until a workman somehow provokes the ship into action. As objects hang telekinetically suspended in the air to the accompaniment of hypnotic pulsing sounds, the workman is quite obviously *possessed* and begins to run through the streets of London in a peculiar spasmodic manner rather like a marionette on strings. Eventually, he finds sanctuary at a church. When Judd and Quatermass arrive to question him, he gives the following account of his experience: "I had to run, to get away. They were coming. I remember it started, and then I could only see them, like you found down there, with eyes and horns. They were alive, jumping, like very fast, and hundreds and hundreds. I knew I was one. Jumping, leaping, oh huge, right up into the sky." "The sky?" Quatermass asks, "what color is it, blue?" No, it seems, the sky was "brown, dark, dark purple." The meaning of this strange report is not long in coming: "What's been uncovered," Quatermass says, "is evil, as ancient and diabolic as anything on record. I think what he gave us just now was a vision of life on Mars five million ago." The workman had evidently been transported to the surface of prehistoric Mars where, to his horror, he found that he was himself a Martian: the man had become-insect.

In order to confirm the workman's unnerving story, Quatermass and Ronay put together a device that will enable the video recording of mental experience. They return to the pit where they hope to document the spacecraft's ability to cause those in proximity to it to become-insect. Neither Ronay nor Quatermass prove ideal experimental subjects, but Judd, in keeping with the tradition that links unconscious mental processes with the female gender, proves especially sensitive to the Martian spacecraft's telepathic emanations. "I can see, I can see," she cries, as mingled expressions of fearful excitement and ecstasy play across her face. After studying the videotape of Judd's visions of jumping and leaping grasshoppers, Quatermass concludes that the scenes she had witnessed of locusts pursuing one another on the plains of Mars suggest a "race purge, a cleansing of the Martian hives. . . . I think we may have witnessed ritual slaughter to preserve a fixed society, to rid it of mutations. . . . That's the way they lived. And it's the way they intended their substitutes on Earth to live." Human aggressivity, xenophobia, and the propensity for social hierarchy are thus that "something else" the Martians implanted in their hominid captives along with high intelligence.

Quatermass's assessment is decisively confirmed when the ship, which is not mechanical but an organism itself and therefore a living embodiment of evolution understood as an infernal power, telepathically reaches out and provokes a "race purge" among the inhabitants of contemporary London. The intended victims are all those who have evolved away from their insect heritage in the direction, ironically, of humanity's ideal image of Man as a rational being. The ship harnesses psychic energy, projecting it into the sky over London in the form of a glowing apparition of the horned god who now presides over the pandemonium below. Neither Judd, who is especially susceptible to the murderous impulses of her Martian heritage, nor Quatermass, who turns out to be half man, half beast, can avoid succumbing. Judd gazes in worshipful ecstasy at the horned god, while Quatermass sedates himself with scotch and chants his name and social position like a protective charm—"My name is Bernard Quatermass, professor of physics, controller of British Experimental Corp., at present engaged in. . . . " Only Ronay, who qualifies as a fully "human" mutation because he is devoid of affect, a purely intellectual being, can defeat the monstrous figure shimmering in the sky over London. In a scene reminiscent of the conclusion to *Attack of the Crab Monsters,* Ronay climbs a nearby construction crane and rides it into the glaring face of the demon, grounding its energy to the earth in a fireball explosion. The most human of all humans thus sacrifices himself in order that his flawed fellow beings might henceforth live free from subjection to their unconscious insect selves. In the closing shot, Quatermass and Judd, positioned on either extreme of the screen and separated by the credits, stand mute, in shock, unwilling or unable to make eye contact with one another as each reflects in solitude upon this traumatic revelation of the demonic element in human nature.

Becoming-Rodent: *My Uncle in America*. Directed by Alain Resnais. Screenplay by Jean Gruault. Based on the works of Henri Laborit. France: Philippe Dussart/Andrea Films, 1980.

> Seeking to dominate in a space we can call the territory is the fundamental basis of all human behavior, though we are not conscious of our motives.

The sobering presentation in *Five Million Years to Earth* of the presence of a horrifyingly "primitive" component in human nature is given a more straightforward scientific exposition in Alain Resnais's *My Uncle in America*. In this film, the life stories of three characters become the object for scientific reflection concerning the biological basis of cultural behavior. The characters' biographies are recounted, first, in brief dossier fashion so as to confer on them the quality of scientific specimens or exhibits, and then, more leisurely, in a conventionally "realistic" narrative style. Periodically throughout the film, scientist and *philosophe* Henri Laborit appears to deliver brief explanations of the evolutionary significance of events as they unfold in the lives of the three characters.

They are, first of all, Jean Le Gall, scion of a provincial bourgeois family, who moves as a young man to Paris hoping to become a famous intellectual. When we pick up his story in the film, he has become a high executive with the National Radio. He is discharged only eighteen months later, however, and in the aftermath of his firing writes a controversial exposé of political infighting at his former workplace. Foregoing his intellectual ambitions, he now becomes a politician and, at the end of the film, is running for elective office.

During the eighteen months of his tenure at the radio, he leaves his wife Arlette and enters into a relationship with Janine Garnier, a young woman whose life story is also related in the film. Garnier comes from a politically militant working-class background. Revolting against what she regards as the claustrophobic moralism of her family, she aspires to the life of a bohemian artist and joins an acting troupe. She meets Jean following her starring performance in a successful Left Bank theatrical production. But the affair with Jean comes to an unhappy, confused, and premature end when Jean's wife Arlette appears before Janine one day and pretends that she has contracted a fatal disease. Filled with compassion for Arlette but also, perhaps, a bit fed up with Jean whose psychosomatic ailments have made him something of a nuisance, Janine precipitates a breakup in order to drive Jean back to his wife who may then spend her few remaining months with the father of her children. At about the same time, Janine changes careers, becoming a rising executive with an international conglomerate that specializes in the textile trade.

This association brings her into contact with the third principal character of the film, René Ragueneau. An apolitical and devoutly Roman Catholic farmer's son, René throws off his father's traditionalism by leaving the farm and becoming an accountant with a small textile firm where he is eventually promoted to the rank of plant manager. When the international conglomerate for which Janine works acquires René's firm, his promotions continue. Unfortunately, the conglomerate elevates him beyond the level of his competence and he is about to be demoted by Janine to the position of manager of a gourmet kitchenwares store when he attempts suicide.

The film ends with René, who has survived his suicide attempt, facing dismissal from his job on the grounds of mental instability, and Janine, having discovered Arlette's duplicity, in despair over her continuing love for the supremely adaptable Jean who has settled once more into marital contentment with Arlette notwithstanding his awareness of her subterfuge.

Two observations are worth making about these rather unexceptional though frequently poignant life stories. First, each of the three figures, in scenes in which they evidently feel oppressed by the run of events, recites the phrase "my uncle in America." The phrase alludes to the proverbial uncle who has gone off to the United States where he has become fabulously wealthy, the familiar fantasy of America, that is, as a land of plenty whose illusory character is underscored by the scenes of urban devastation in the South Bronx with which the film closes. "My uncle in America" expresses a desire to be done with struggle and conflict, to dwell in peace and security with every want abundantly supplied; a desire, finally, to rise above the world to a position of invulnerability and perfect fulfillment: in the words of an old Talking Heads' song, "Heaven is a place where nothing ever happens." Second, all three characters, as adults, contradict the values and principles dear to them in their youth: Jean, the would-be intellectual removed from the chicanery of public life, winds up a politician who obsequiously curries favor with the established powers; Janine, the bohemian artist, is rising quickly in the business world; and the hapless René has mortally sinned against his faith by attempting suicide. The three characters' conscious values, first principles, and ideological commitments are thus of little value in predicting their actual conduct.

The explanation both for the recurrent fantasy of consummate satisfaction and for the inconsistency of character according to which Jean, Janine, and René each betray their respective ideal self-images is implicitly conveyed in the course of Professor Laborit's remarks. According to Laborit, the maintaining of a condition of biological equilibrium can be said to orient the vital processes of every organism, no matter whether plant or animal. "A being's only reason for being," he says, "*is* being. That is to say, it must maintain its structure. It must stay alive. Otherwise, there is no being." Unlike plants, animals pursue this end by means of a nervous system that "permits action upon, and within, the environment. And always for the same reason, to insure survival." Further, in the case of animals sufficiently complex for their nervous system to culminate in that complex and intricate neural network known as the brain, it is this organ that is responsible for initiating and directing the organism's self-preservative behavior. Turning to the specific case of the human brain, Laborit observes that the brain preserves its own history in its present constitution. The human brain, that is, is not unitary but "triune" in structure, its three parts representing three successive moments of evolutionary innovation.

It turns out that one portion of the brain—an archaic structure fifty million or more years older than those that sit atop it—is especially consequential with respect to the human organism's efforts to maintain biological equilibrium. Laborit borrows here from the research of Paul MacLean, who has dubbed this structure the "R-complex" or "reptilian brain," and bluntly sums up its significance: the R-complex insures that "seeking to dominate in a space we can call the territory is the fundamental basis of all human behavior, though we are not conscious of our motives."

But if "our drives are still primitive, coming from the reptilian brain," these drives are mediated in complex ways by the cultural conditioning that especially affects, Laborit says, the second level of the brain in MacLean's scheme. That is, the characteristic affective orientation toward the world that distinguishes one personality from another is largely the result of environmental factors which from a very early age shape and contour the "limbic system" or "affective brain." The limbic system, wedged between the R-complex and the cerebral cortex (locus of complex cognitive functions including language), is found in humans and all other mammals but is largely absent in reptiles. It deals, Richard Restak remarks, "with the emotional feelings that guide behavior." When portions of the limbic system are removed, mammals lose many behaviors—for instance, playfulness and nurturing—that are precisely characteristic of the mammalian order. Whatever behaviors remain are apt to resemble those of reptiles (see Restak 1984: 136–37). Laborit dissents slightly from this characterization of the second layer as the "affective brain," preferring to call it "the memory brain": "With no memory of what is pleasant or unpleasant, there's no question of being happy, sad, anguished, nor of being angry, or in love. We could almost say that a living creature is a memory that acts."

My Uncle in America provides many examples of the cultural conditioning that justifies this characterization of human beings as memories that act: scenes, for instance, of children being encouraged by parental figures to internalize the "bric-a-brac of value judgments, prejudices, and platitudes" typical of their respective social milieus. But the most striking (and amusing) instances of imprinting are presented in the form of clips from old black-and-white films that appear at crisis moments in the narrative. What the clips reveal are the unconscious ego-ideals of each character. Internalized icons of the silver screen provide reference points for the characters as they attempt to negotiate life's various challenges, stereotypical stances toward the world that orient, guide, and sustain the illusion of autonomous agency.

The point here is that "dominance" and "territory" are empty terms until they are filled with cultural contents. "There is no proprietary instinct," Laborit explains, "Nor is there an instinct to dominate. The individual's nervous system has simply learned the necessity of keeping for the individual's own use an object or person that is also desired or coveted by another being. And he has also learned by experience that in the competition to keep that object or that person for himself, he must dominate." Laborit can thus sum up by saying that of the three structures comprising the human brain "the first two function unconsciously, beneath our level of awareness: drives, cultural automatisms. The third furnishes an explanatory language which gives reasons, excuses, alibis, for the unconscious workings of the first two." As in Freud, the ego is not the master in its own house: human behavior is propelled by unconscious drives and cultural automatisms "masked" by conscious beliefs that only serve "to hide the cause of dominance."

As proof for these remarks about human nature, Laborit presents representative instances of "flight," "fight," and "anguish" behaviors as manifested in laboratory rats that are then correlated with the respective behaviors of Jean, Janine, and René. The first example concerns a rat that is placed in a cage whose floor is intermittently electrified. The rat learns that the sound of a bell announces the imminence of an unpleasant electric shock and accordingly flees through a doorway to the other side of the cage which is not electrified. When the bell sounds again, this time signaling the imminent electrification of its present location, the rat returns to its starting place. The rat remains "in perfect health. . . . He has maintained his biological equilibrium." Similarly, when Jean realizes that his rise in the intellectual hierarchy of Parisian society is hindered by his association with Arlette, who is provincial and poorly educated, he "flees" her for Janine. When he loses his position at the National Radio and Janine provokes a break-up, he "flees" Janine, returns to Arlette, and thereafter decides to become a politician. Though he may seem shallow or even hypocritical, Jean's behavior is in fact perfectly consistent: when he cannot "dominate" one "territory," he moves to another; when one of the cultural clichés he has internalized—the social prestige that accompanies intellectuality—yields no reward in his own case, he shifts to another—the power the politician may wield to improve the world. And all the while in the back of his mind he fantasizes about his mythical "uncle in America," whom he identifies with the hero of a comic book from his youth: the "Gold King, Samuel Knight, orphan and millionaire . . . in the true American way. . . . "

In the second example, two rats are placed in a cage, the floor is intermittently electrified, and the escape route is closed off. The rats respond to this predicament by fighting one another when the floor is electrified. Their behavior in no way addresses the cause of their distress, but the simple fact of taking action—"a nervous system is meant to act"—enables them to remain healthy, their blood pressure normal, their coats sleek. Similarly, at the end of the film, Janine confronts Jean and attacks him when it becomes apparent that she cannot win him back. Her behavior does nothing to alter the situation, but it prevents her from succumbing to "anguish," or the despair that accompanies the "impossibility of dominating a situation." Finally, in the third experiment, a rat is placed in a cage with neither the possibility of escape nor an adversary against which to vent its distress. The intermittent electrification has disastrous consequences. The rat exhibits anguish behavior: it becomes depressed, develops ulcers, high blood pressure, and so on. In Laborit's terms, it aggresses its own body. In like fashion, psychosomatic ailments and mental illness culminating in a suicide attempt afflict René as his career comes unraveled.

This is hardly a comforting or ennobling depiction of the human situation. Yet the mood of the film, much of the time, is surprisingly buoyant and high-spirited. As one way of explaining this discrepancy, consider the following: after the dossiers on the three principal characters have been presented at the beginning of the film, a fourth dossier is added, this one concerning Henri Laborit himself, whose dossier chronicles his many scientific accomplishments and professional accolades. This presentation of Laborit's achievements does not, however, have the effect of credentialling him as the voice of truth. Rather, by positioning him as yet another character in the film's narrative, the film reminds its audience of the cultural and historical situatedness of scientific discourse. At the same time that it takes Laborit's views seriously as a valuable albeit unflattering perspective, the film underscores their perspectival character with the consequent implication that "something else," as Quatermass might put it, has been left out of account in Laborit's rendering of the human situation.

In the sequences presenting Laborit's views on the biological foundations of human social life with reference to the responses of laboratory rats to reward and punishment scenarios, the equation between rat behavior and human behavior is augmented by reprises of key scenes in the narrative in the course of which, the second time through, the actors themselves don white rat costumes. Miniatures of some of the sets are also shown—Jean and Arlette's Paris apartment, for instance—populated by real white rats. The effect of these literal enactments of becoming-rodent—along with other scenes in which human conduct is equated with the respective behaviors of crabs, fish, tortoises, and wild pigs—undermines the air of gravity and sobriety one might expect given the ostensive message of the film. Though the film is undoubtedly poignant in passages, it does not fill the viewer with a disabling sensation of anguish at this deflation of human dignity. Nor, I think, is the concluding mood one of satisfaction at having transcended, and thus dominated, the human situation by regarding it from a commanding overlook of omniscience. Instead, a spirit of carnivalesque festivity prevails. Laborit's perspective on human nature has been narrativized not as tragedy but as *farce.* In other words, the spectacle of becoming-rodent introduces into the experience of viewing the film an element of hilarity and even "exhilaration" (to borrow *Attack of the Crab Monsters*' favorite term for becoming-crustacean) that can't be explained with reference to Laborit's construction of human nature.

There is a similar untheorized component in *Five Million Years to Earth.* In this case, the film ostensively reflects on the origins of violence and hierarchy in humanity's horrific subjection to evolutionary process. But the introduction of the Gothic motif of "satanism" complicates the viewer's response to the film. The horned god is, of course, not only Satan, the embodiment of everything evil, but Dionysus as well, the god of libidinal excess. In other words, a covert eroticization of becoming-insect accompanies the horror entailed in realizing human participation in "nature red in tooth and claw." As this essay moves now to consider its final example of evolutionist cinema, I'm going to focus precisely on this aura of mingled hilarity, exhilaration, and erotic excitement that surrounds the respective scenarios of becoming-crustacean in *Attack of the Crab Monsters,* becoming-insect in *Five Million Years to Earth,* and becoming-rodent in *My Uncle in America.*

Becoming-Fungus: *Attack of the Mushroom People.* Directed by Inoshiro Honda and Eiji Tsuburaya. Screenplay by Takeshi Kimura. Adapted from "The Voice in the Night" (1907) by William Hope Hodgson. Japan: Toho, 1963.

> Yeah. I ate mushrooms. Now you know. I read in a book a long time ago that the Mexicans used to eat them in order to enhance their perceptions and get a sense of well-being. . . . Japanese legends mention laughing mushrooms, so I'm in good company. The people who went out to gather the mushrooms danced in high spirits in the mountains and were in touch with the infinite. But [these] Matango [mushrooms], according to your understanding, leave a person no longer human. Well, that's fine with me.

Inoshiro Honda and Eiji Tsuburaya's *Attack of the Mushroom People* relates the adventures of a group of weekend vacationers on board a yacht who encounter a typhoon and are shipwrecked on an uncharted island. The five men and two women comprise a social cross-section: a skipper and his mate, both in the employ of the wealthy entrepeneur who owns the yacht; a nightclub chanteuse who is the entrepeneur's mistress; a writer of detective fiction; a psychology professor and his ingenue graduate student. At the beginning of the voyage, spirits run high as Yoshida, the writer, proposes a toast to their temporary reprieve from the hustle-bustle of social life: they are free, he says, from all the problems of Man. But such a release from worldly care is illusory, Meimi, the chanteuse, counters when she ominously replies "Aren't we part of humanity?" In other words, the social ills and antagonisms Yoshida hopes to have escaped are not so easily eluded. The yacht's complement are already in conflict along a number of axes: the men condescend to the women and are themselves divided against one another according to class position, educational attainment, and so on, as when the skipper and his mate, alone on deck, denounce their social superiors as "reckless children" and "parasites." The film's seven principal characters thus constitute a microcosm of the great world they have left behind where Laborit's hierarchies of dominance are amply instanced in flashbacks that present, for instance, tycoons in a Tokyo nightclub reveling in their eminence. As one tycoon boasts that "I hire people to think for me, then I just use their thoughts," the other complacently replies, "that's right, that's how civilization progresses, by borrowing, each generation takes an idea and improves on it." Meanwhile, they enjoy a parade of female bodies offered for their perusal in a Folies Bergères-style revue.

The aggressive tensions that have already spoiled the holiday excursion of the yacht's crew and passengers are temporarily set aside in the immediate aftermath of their shipwreck in the interest of collective survival. The seven set out together to search for food and soon come upon a much larger vessel, beached and evidently deserted. The ship—which the castaways soon occupy as a refuge from the monsoon-like weather—turns out to be a research vessel apparently sent to the island to investigate the effects of radiation from atomic tests on the local ecosystem. Now derelict, the vessel is in a state of considerable disrepair: a disgusting, slime-covered fungus everywhere coats its interior. As they rummage through the cabins in search of food, the castaways open a large specimen crate in which they find a giant mushroom—evidently a radiation-induced mutation—that has been labeled "Matango." The skipper remarks hopefully: "We're in luck if this thing is edible," but such does not appear to be the case. The psychologist Morrei, after studying the ship's log and assorted scientific documents, reports that not only is the island virtually barren of edible plants and animals but the Matango mushroom must be avoided because it "damages nerve tissue."

The little society of castaways soon realizes that the warning the ship's previous occupants have left behind them concerning the mushroom's toxic effects is in reality a euphemism for a far more disturbing consequence of ingesting Matango. The mushroom has mutagenic properties: those who eat Matango become-fungus themselves, as the castaways discover one dark night when they come face to face not with a ghost but with a still living member of the research vessel's former crew who, like his shipmates, was made so desperate by starvation that he dined on Matango and thereby lost his humanity. His body, though vaguely humanoid in outline, is covered with fungus-like protuberances and resembles far more the mushrooms he has consumed than it does a truly human form.

But becoming-fungus may not, after all, amount to the proverbial fate worse than death, especially when compared with the dissension and eventually overt violence that afflict the castaways as they grow increasingly desperate from starvation themselves. However repulsive the prospect of a mushroom-like existence may at first glance seem, the film suggests that this destiny can in fact be understood as equivalent to a multiplication and enhancement of life's possibilities. Certainly this is Yoshida's view when he storms off into the jungle, saying: "I'm going after that thing that came on board and frightened everybody . . . I certainly hope it is a man because then I'll be able to talk to him. I'll be very interested to hear what he says. A man thinks strange things when he's out of his mind. His reality may be more fascinating than ours, and it wouldn't matter what he ate." With reference to what Yoshida discovers when he does catch up with the mushroom man, that last portion of his outburst ought to be revised. The mushroom man's reality is more fascinating than the castaways' *because* of what he ate: the Matango are "magic mushrooms." The speech Yoshida delivers upon his return to the ship—in which he alludes to Native American peyote rituals and to similar customs in Japan—confirms he had already suspected as much. And he evidently considers the opportunity to consume magic mushrooms that put one "in touch with the infinite" well worth the price, in the case of the mutagenic Matango mushroom, of his humanity.

Yoshida has hardly proven an admirable human being himself—when he returns from the jungle he kills the mate while trying to kidnap the ingenue Akiko and is later expelled from the ship with Meimi when the two of them conspire to cause further trouble—but the film's explicit argument is, after all, that aggression, competitive struggle, and the will to ascend to the summit of the social pecking order, are *the* quintessentially human attributes, traits definitive of "human nature." In becoming-fungus, Yoshida does himself a favor. Although the mushroom people can be criticized for, so to speak, over-enthusiastically proselytizing their way of life when they swarm over the ship in an attempt to carry off Morrei and Akiko, the last remaining humans on the island, little indication is given that dominance behavior constitutes a significant dimension of their social life.

Morrei, the sole survivor to make it back to civilization, says as much at the very end of the film: "People in cities are cruel aren't they? . . . I would be happier to live on that island than in this city," happier to have become-fungus alongside Akiko, who herself finally succumbed to temptation and ate Matango, than return to a society that has fended off the disquieting implications of his testimony by confining him in a psychiatric ward. The voice of scientific rationalism at the beginning of his adventure—"Man's . . . reason must grow stronger if we are to continue to progress"—Morrei has lost his faith in the Enlightenment grand narratives of universal knowledge and the inherently progressive character of cultural history. "Progress," he now believes, is merely warfare by another means, a sublimation of overt dominance behavior.

The mushroom people, who laugh uproariously and derisively at anyone who struggles to remain merely human, would no doubt agree with this assessment. The mushrooms they've consumed have transported them to a very different reality. That is, the Matango are "magic mushrooms" in two respects: they have hallucinogenic properties and they are aphrodisiacs. Becoming-fungus thus entails entering a realm of imaginative and libidinal excess. It is no accident that the first two castaways to succumb to the temptation of the mushrooms are Yoshida, the writer, and Meimi, the chanteuse, both of whom are more libidinally opportunistic than their companions. The first two "victims" of becoming-fungus are precisely the least invested in maintaining a settled egocentric psychological structure, the most receptive to the wayward impulses of unconscious drive.

It comes as something of a surprise, then, to witness that in their hallucinations the mushroom people return to civilization, to the great world that had previously been shown in flashback to be permeated with aggressive tensions and dominance hierarchies. But what is now foregrounded is the *fantastic* character of the multifarious amusements and pastimes of Tokyo nightlife, the inventive fecundity now shown to have resided all along in human culture. From this perspective, the mushroom people ought to be interpreted not as figuring transcendence to an unheard-of beyond but as tropes for hitherto abominated capacities of the human body itself, capacities at least potentially in conflict both with the everyday tactical maneuvering for advantage instanced throughout the film and with such strategic consolidations of dominance as Modernity's normalizing definition of "Man."

From the perspective of evolutionary biology, *Attack of Mushroom People* can be said to thematize two characteristics of the human body left unaddressed in Laborit's account. First of all, the brain's powers of invention are far in excess of survival requirements, a characteristic tracing to its "neotenous" evolutionary origin. Neoteny, Steven Stanley says, is "a kind of juvenilization, whereby embryonic or youthful traits of our ancestors have been shifted to the adult stage of our development" (1981: 127). In the case of humans, neoteny accounts for the physical resemblance of adult humans to immature chimpanzees and, in Stephen Jay Gould's words, for the development of a large cerebral cortex "through the retention of rapid fetal growth rates" (1977: 77). Neotenous evolution has thus led to an enhancement of learning and play behaviors as instanced in the remarkably various array of cultural practices that humans have invented.

The second point to be made here concerns an odd fact about human sexuality: instead of seasonal or periodic sexuality that is timed to coincide with the possibility of conception, humans are capable of sexual response at any time. Human sexuality, in other words, is substantially de-coupled from the exigencies of reproduction with the result, as psychoanalysis has always maintained and evolutionary theory supports, that desiring energy can invest a broad range of activities and objects through which gratification is obtained: virtually any behavior can be eroticized, virtually anything at all can become an object of desire. Echoing evolutionary theory, Deleuze and Guattari put it this way: "Sexuality is not a means in the service of generation; rather, the generation of bodies is in the service of sexuality as an autoproduction of the unconscious" (1977: 108). Humans are thus constitutionally perverse; their desires are errant, wandering. While any object may be so inflated with erotic significance that it imposes a ritualized pattern of behavior as the epitome of desirability itself, there is also the possibility of a roving desire that is never more than contingently bound for a particular destination.

What thus emerges in the monstrous becomings of evolutionist cinema is a portrait of a body not merely composite but actively at odds with itself. The same body that is horrified by the discovery of the *menagerie within* is also erotically thrilled at the prospect of multiplicitous destinies. A peculiar body, then: at once "human" in its longing for identity and "posthuman" in the exhilaration produced in it by alterity; a body at once prone to fantasies of ultimacy—as "human"—and to fantasies of becoming—as "proteus." An ironic body, too: just as the R-complex or reptilian brain is the motive force propelling the construction and adherence to a specifically "human" identity, so polymorphous sexuality and the limitless capacity of the cerebral cortex for imaginative play, traits arguably specific to the human assemblage, ensue in the production of indeterminately many scenarios of monstrous metamorphosis. But for all that, a body that need not feel itself at an impasse, so long as it remains undismayed by the fact that Fortune who rules this world decrees Sisyphean fates for all.

The monstrous becomings of evolutionist cinema continue a cultural project that was perhaps first broached in a concerted way in the heyday of the dissident surrealists: recall the libidinal economy of unconditional expenditure that Bataille thought he detected in potlatch rituals, or Artaud's body-without-organs—"When you will have made [Man] a body without organs, then you will have delivered him from all his automatic reactions and restored him to his true freedom"—or the collages of Max Ernst about which Margot Norris has brilliantly written: "Ernst's body parts, mutilated bodies, hybridized creatures, and otherwise distorted forms betray an intention to challenge the metaphysical framework that gives rise to concepts of wholeness, of the unity of the body, of the expected relations of parts to whole" (1985: 146). They practically instantiate Deleuze and Guattari's concept of becoming-other. Their respective monstrous becomings—becoming-crustacean, becoming-insect, becoming-rodent, becoming-fungus—"molecularize" the "molar aggregate" of "Man," or submit the "plane of organization" that organizes a body into a unified functioning in the service of the ego to the "plane of immanence" of the body-without-organs, thereby achieving a bodily "nomadism" such that the body may henceforth be understood as an archive of possibilities whose behavior resembles what complex dynamicists call a "strange attractor": an irregularly recurrent, self-similar but infinitely various system whose ceaseless mutations, in the absence of a linear trajectory or goal, transpire according to the strictly additive principle—"and . . . and . . . and . . . "—of schizo-nomadic picaresque (Deleuze and Guattari, 1987: 232–309). Notwithstanding their occasional humanist nostalgia, the monstrous becomings of evolutionist cinema make a happily delirious contribution to a task defined by Deleuze as follows: "to make the body a power which is not reducible to the organism, to make thought a power which is not reducible to consciousness . . . " (Deleuze and Parnet, 1987: 62).

Works Cited

Deleuze, Gilles, and Felix Guattari. *Anti-Oedipus: Capitalism and Schizophrenia.* Trans. Robert Hurley, Robert Seem, and Helen R. Lane. New York: Viking, 1977.

———. *A Thousand Plateaus: Capitalism and Schizophrenia.* Trans. Brian Massumi. Minneapolis: University of Minnesota Press, 1987.

Deleuze, Gilles, and Claire Parnet. *Dialogues.* Trans. Hugh Tomlinson and Barbara Habberjam. New York: Columbia University Press, 1987.

Douglas, Mary. *Purity and Danger.* London: Routledge, 1969.

Gould, Stephen Jay. *Ever since Darwin: Reflections on Natural History.* New York: Norton, 1977.

Margulis, Lynn, and Dorion Sagan. *Mystery Dance: On the Evolution of Human Sexuality.* New York: Summit, 1991.

Norris, Margot. *Beasts of the Modern Imagination: Darwin, Nietzsche, Kafka, Ernst, and Lawrence.* Baltimore: Johns Hopkins University Press, 1985.

Restak, Richard M. *The Brain.* New York: Bantam, 1984.

Stanley, Steven M. *The New Evolutionary Timetable: Fossils, Genes, and the Origin of Species.* New York: Basic Books, 1981.

Thomas, Lewis. *The Lives of a Cell: Notes of a Biology Watcher.* New York: Bantam, 1974.

Contributors

Kathy Acker's twelve novels include *Great Expectations, Blood and Guts in High School, Kathy Goes to Haiti,* and *My Mother: Demonology.* She also wrote the screenplay for the film *Variety* and is the author of the play *Lulu Unchained.*

Alexandra Chasin is Assistant Professor of English at Boston College. She studies, teaches, and writes about material culture, feminist theory, popular culture, and the U.S. in the twentieth century.

Camilla Griggers teaches feminist cultural studies, gender studies, and media theory in the Department of English's Literary and Cultural Theory Program at Carnegie Mellon University. Her book *Becoming-Women* is forthcoming from the University of Minnesota Press's Theory Out of Bounds series. She is coproducer of two pedagogical videos, *Discourse/Intercourse* and *Hirohito's Funeral.*

Judith Halberstam is Assistant Professor of Literature at the University of California, San Diego, where she teaches film, literature, and queer theory. She has a book forthcoming from Duke University Press called *Skin Shows: Gothic Horror and the Technology of Monsters.*

Kelly Hurley is Assistant Professor of English at the University of Colorado at Boulder. She has published articles on nineteenth-century degeneration theory and on late-Victorian Gothic fiction, and is completing a book entitled *Gothic Embodiments of the British Fin de Siècle.*

Ira Livingston is Assistant Professor of English at the State University of New York at Stony Brook. He is completing a book about Romanticism and postmodernity.

Carol Mason is finishing her doctoral work in English at the University of Minnesota.

Paula Rabinowitz teaches cultural studies at the University of Minnesota. She is the author of *They Must Be Represented: The Politics of Documentary* and *Labor and Desire: Women's Revolutionary Fiction in Depression America.* She is also coeditor of *Writing Red: An Anthology of American Women Writers, 1929–1941.*

Roddey Reid is Associate Professor of French Literature at the University of California, San Diego. He is the author of *Families in Jeopardy: Regulating the Social Body in France, 1750–1910* and editor of *Located Knowledge,* a special issue of *Configurations: A Journal of Literature, Science, and Technology.*

Steven Shaviro is the author of *Passion and Excess: Blanchot, Bataille, and Literary Theory* and *The Cinematic Body.*

Susan M. Squier is Julia Gregg Brill Professor of English and Women's Studies at the Pennsylvania State University. She is the author of *Babies in Bottles: Twentieth Century Visions of Reproductive Technology* and *Virginia Woolf and London: The Sexual Politics of the City;* she is editor of *Women Writers and the City: Essays in Feminist Literary Criticism* and coeditor of *Arms and the Woman: War, Gender, and Literary Representation.*

Allucquere Rosanne Stone is Assistant Professor of Radio, TV, and Film at the University of Texas at Austin. Her book *The War of Desire and Technology at the Close of the Mechanical Age* is forthcoming from MIT Press. She is currently working on her second book, *The Gaze of the Vampire* and an anthology called *Dangerous Bodies.*

Jennifer Terry teaches courses in the cultural studies of science program at Ohio State University. She is coeditor of *Deviant Bodies* and is completing a book on the history of scientific research on homosexuality.

Eric White teaches twentieth-century literary cultural studies in the English Department at the University of Colorado at Boulder. He is the author of *Kaironomia: On the Will-to-Invent* and of essays on the rhetoric of scientific discourse in such fields as chaos theory, contemporary cosmology, and evolutionary biology.

Index